Sustainable Development Goals Series

The **Sustainable Development Goals Series** is Springer Nature's inaugural cross-imprint book series that addresses and supports the United Nations' seventeen Sustainable Development Goals. The series fosters comprehensive research focused on these global targets and endeavours to address some of society's greatest grand challenges. The SDGs are inherently multidisciplinary, and they bring people working across different fields together and working towards a common goal. In this spirit, the Sustainable Development Goals series is the first at Springer Nature to publish books under both the Springer and Palgrave Macmillan imprints, bringing the strengths of our imprints together.

The Sustainable Development Goals Series is organized into eighteen subseries: one subseries based around each of the seventeen respective Sustainable Development Goals, and an eighteenth subseries, "Connecting the Goals," which serves as a home for volumes addressing multiple goals or studying the SDGs as a whole. Each subseries is guided by an expert Subseries Advisor with years or decades of experience studying and addressing core components of their respective Goal.

The SDG Series has a remit as broad as the SDGs themselves, and contributions are welcome from scientists, academics, policymakers, and researchers working in fields related to any of the seventeen goals. If you are interested in contributing a monograph or curated volume to the series, please contact the Publishers: Zachary Romano [Springer; zachary.romano@springer.com] and Rachael Ballard [Palgrave Macmillan; rachael.ballard@palgrave.com].

More information about this series at https://link.springer.com/bookseries/15486

Christopher B. Barrett · Tim Benton · Jessica Fanzo · Mario Herrero · Rebecca J. Nelson · Elizabeth Bageant · Edward Buckler · Karen Cooper · Isabella Culotta · Shenggen Fan · Rikin Gandhi · Steven James · Mark Kahn · Laté Lawson-Lartego · Jiali Liu · Quinn Marshall · Daniel Mason-D'Croz · Alexander Mathys · Cynthia Mathys · Veronica Mazariegos-Anastassiou · Alesha Miller · Kamakhya Misra · Andrew Mude · Jianbo Shen · Lindiwe Majele Sibanda · Claire Song · Roy Steiner · Philip Thornton · Stephen Wood

Socio-Technical Innovation Bundles for Agri-Food Systems Transformation

palgrave
macmillan

Christopher B. Barrett
Cornell University
Ithaca, NY, USA

Tim Benton
University of Leeds
Leeds, UK

Jessica Fanzo
Johns Hopkins University
Baltimore, MD, USA

Mario Herrero
Cornell University
Ithaca, NY, USA

Additional Authors *see next page*

ISSN 2523-3084 ISSN 2523-3092 (electronic)
Sustainable Development Goals Series
ISBN 978-3-030-88801-5 ISBN 978-3-030-88802-2 (eBook)
https://doi.org/10.1007/978-3-030-88802-2

Cover Credit: ALAMY B093NN

This Palgrave Macmillan imprint is published by the registered company Springer Nature Switzerland AG
The registered company address is: Gewerbestrasse 11, 6330 Cham, Switzerland

Additional Authors

Rebecca J. Nelson
Cornell University
Ithaca, NY, USA

Edward Buckler
US Department of Agriculture
Agricultural Research Service
Ithaca, NY, USA

Isabella Culotta
Cornell University
Ithaca, NY, USA

Rikin Gandhi
Digital Green
San Francisco, CA, USA

Mark Kahn
Omnivore
Mumbai, India

Jiali Liu
Cornell University
Ithaca, NY, USA

Daniel Mason-D'Croz
Cornell University
Ithaca, NY, USA

Cynthia Mathys
Cornell University
Ithaca, NY, USA

Alesha Miller
Digital Green
San Francisco, CA, USA

Andrew Mude
African Development Bank
Abidjan, Côte d'Ivoire

Lindiwe Majele Sibanda
University of Pretoria
Pretoria, South Africa

Roy Steiner
Rockefeller Foundation
New York, NY, USA

Stephen Wood
The Nature Conservancy
New Haven, CT, USA

Elizabeth Bageant
Cornell University
Ithaca, NY, USA

Karen Cooper
Nestlé
Vevey, Switzerland

Shenggen Fan
China Agricultural University
Beijing, China

Steven James
PepsiCo
Dallas, TX, USA

Laté Lawson-Lartego
Oxfam America
Washington, DC, USA

Quinn Marshall
Johns Hopkins University
Baltimore, MD, USA

Alexander Mathys
ETH Zürich
Zürich, Switzerland

Veronica Mazariegos-Anastassiou
Brisa de Año Ranch
Pescadero, CA, USA

Kamakhya Misra
Cornell University
Ithaca, NY, USA

Jianbo Shen
China Agricultural University
Beijing, China

Claire Song
Cornell University
Ithaca, NY, USA

Philip Thornton
International Livestock Research Institute
Edinburgh, UK

Foreword

The seventeen UN Sustainable Development Goals (SDGs) for 2030 galvanized attention on the most urgent global needs. One of the strengths of the SDGs is that collectively they reflect the dependence of human thriving on social equity and the health of the environment, including climate and biodiversity on land and under water. One of their shortcomings is that if addressed one at a time, the likelihood of trade-offs among them may not be recognized, resulting in failure overall. If, for example, Zero Hunger (Goal 2) is pursued single-mindedly with only short-term goals in mind, and using traditional Green Revolution-style agriculture, the likelihood of achieving Responsible Production and Consumption (Goal 12), Life Below Water (Goal 14), and Life on Land (Goal 15) will be dramatically diminished.

The expert panel tackled this conundrum head-on with respect to the global systems that produce and distribute food. The panel's rigorous synthesis and analysis of existing research leads compellingly to multiple actionable recommendations that, if adopted, would simultaneously lead to healthy and nutritious diets, equitable and inclusive value chains, resilience to shocks and stressors, and climate and environmental sustainability. The panel refers to this set of goals as HERS (Healthy, Equitable, Resilient, Sustainable), implicitly emphasizing the importance of SDG 5: Gender Equality.

The Cornell Atkinson Center for Sustainability is pleased to have partnered with *Nature Sustainability* to convene the international panel of 23 experts who, along with several coauthors, produced this report. The robustness of the analysis and recommendations is the result of the transdisciplinary spirit that animated the group, which included representatives from many research disciplines, and from multiple economic sectors and kinds of organizations, including universities, the food and financial industries, environmental and humanitarian organizations, national and multilateral government organizations, and a philanthropic foundation. The organizations represented included The Nature Conservancy and OXFAM, both strategic partners of Cornell Atkinson. The panel embodied co-creation, one of the key recommendations the panel makes for the needed innovations in agri-food systems, and one of the hallmarks of all the work we do at Cornell Atkinson.

Equally important, this panel eschewed simple, one-size fits all, recommendations that often lack sensitivity to the roles of culture and biophysical constraints that differ markedly across the planet. That means that the analysis and recommendations are more complex in concept, but not necessarily more difficult to implement. In fact, it is quite possible that the kinds of recommendations made here are much more likely to be adopted because of the recognition that the appropriateness of any given agri-food innovation is contingent on culture and geography at least. This does require a conceptual paradigm shift.

The agri-food systems that developed from the Green Revolution, and were put into practice in strikingly similar ways across the Global North and much of Asia, are like a tragic hero in ancient Greek literature. Bad things befall a person of exceptionally high moral character and accomplishment because of a wrong choice or even a fatal flaw. Our current agri-food systems have delivered an astonishing increase in global food production since the 1950s and consequently a terrific decrease in global hunger. Simultaneously, however, they have also led to many negative externalities, including increased obesity and diet-related noncommunicable diseases, poor working conditions and inadequate income for many laborers throughout the value chain, declining water quality, and loss of terrestrial and aquatic habitat and biodiversity. Our food is cheap because it does not reflect the true human welfare and environmental costs of producing food in this way. The appropriate transformation is more complex, however, than simply increasing food prices, which would instantly increase global food insecurity.

The report makes clear that continuing on the current agri-food system trajectory requires multiple substantial changes. Without them, the already large negative side effects of current food production and distribution will become overwhelming in the face of growing challenges. Human population is still increasing, especially in Sub-Saharan Africa, and will not peak globally until the 2060s. Humans are increasingly concentrated in cities, and removed from the sites of traditional food production—a trajectory that contributes to the growing socio-politically unstable divide between rural and urban populations. Global per capita consumption of foods with large environmental footprints, including meat, is growing. Climate change, the COVID-19 pandemic, and various market forces have led to increasing volatility of food prices. All these forces decrease or indicate a loss of socio-economic resilience in the agri-food system.

Therefore, as the report's authors explain, a different approach is needed in the coming decades. To transition to an agri-food system that is HERS, no single approach will suffice. Rather the report recommends a middle path between globally scaled practices, such as we now have, and wholly local practices that would suffer the inefficiencies of overly small scale. They describe the middle path as requiring the development of regionally fit-for-purpose socio-technical innovation bundles that are designed with trade-offs in mind to maximize the overall increase in human welfare over long time frames.

The "socio" part of the innovation bundles would include reforms of institutions and cultural practices, including changes in government policies in many countries. For example, net societal benefits would increase by carefully transferring expenditures on agricultural subsidies that prop up current agri-food systems to programs that reduce systemic risk and foster social protection. New policies would need to enable or incentivize technical innovation. The "technical" part of the innovation bundles would include new digital platforms to increase civic engagement and decentralize power in value chains, new financing structures and products to increase the flow of private capital into an increased diversity of agri-food system practices and products, and, of course, more innovation in plant breeding, agronomic, and food manufacturing practices that will increase production of nutritious foods with a lower water and land footprint.

The "socio-" and "technical" innovations must be bundled for at least three reasons. First, the "socio-" and "technical" are not independent in their origins or impacts. Policy can enable or even drive innovations in research and technology, which in turn can necessitate innovation in policy. Second, rarely will one innovation be appropriate at global scale. Instead different combinations of institutional and technological innovations will be appropriate in Europe, North America, Asia and Sub-Saharan Africa. Third, tradeoffs inherent with every innovation foster opposition to innovation; every innovation causes someone's ox to be gored. Bundling innovations can address tradeoffs to bring all parties along. The current trends reviewed in this report make crystal clear that it is most urgent that the development of socio-technical innovation bundles be focused on post-farmgate institutions and practices, especially in Sub-Saharan Africa.

The challenges are great, with opportunities to match. We at Cornell Atkinson are pleased to have supported this panel and its report. I hope that this report will produce further innovations in research and ultimately in the production of a more sustainability-focused virtuous loop between institutions and practices in the place-appropriate production and distribution of food.

David M. Lodge
Francis J. DiSalvo Director, Cornell Atkinson
Center for Sustainability
Professor, Department of Ecology
and Evolutionary Biology
Cornell University
Ithaca, NY, USA

Acknowledgements

We thank Phil Campbell, Sara Farley, Xiaoqiang Jiao, Charlotte Pedersen, Jaron Porciello, Roseline Remans, and Fusuo Zhang for valuable input and guidance; Alex Goddard, Julia Hans, Christine Johnson and William Stafstrom for helpful research assistance; Lizzy Barrett for graphic design; and Jackie Swift for expert copy editing. We thank John Antle, Ken Giller, Molly Jahn, Meha Jain, David Lobell, Andrew McDonald, and Cheryl Palm for constructive comments as external expert peer reviewers of an earlier draft. We also thank Patrick Beary, Kurt Fritjofson, David Lodge, John McKain, and Sara Levin Stevenson for their support to the panel. We acknowledge, in particular, the innovations inventory and the body of analyses supplied by Australia's Commonwealth Scientific and Industrial Research Organisation (CSIRO)'s Wild Futures Project (Herrero et al. 2020, 2021), on which we build. The authors write in their personal capacities. None of the opinions reflected in this report represent any organization.

Praise for *Socio-Technical Innovation Bundles for Agri-Food Systems Transformation*

"This seminal report clearly lays out the imperative to transform our agri-food systems to prioritize healthy and prosperous people and planet. This non-negotiable clarion call is most urgent in Africa, and strongly resonates with the African Development Bank's Feed Africa strategy to transform Africa's agri-food systems to align with the "HERS" goals championed in this report: A food system that optimizes for Healthy, nutritious diets; Equity and inclusivity for women and youth; Resilience to climate change and other stressors; and environmentally Sustainable growth. The recommendation to bundle tailored scientific innovation within enabling institutional frameworks offers a compelling directive for the design of impactful agri-food system investments. An essential guide for any committed agent of agri-food systems' change in Africa and beyond."

—Akinwumi A. Adesina, Ph.D., *President, African Development Bank Group, 2017 World Food Prize Laureate*

"The word "innovation" is, for many, synonymous with technological silver bullets. If this is you, reader, read on. You will be convinced that innovation needs to come in bundles: in technology but critically also in agency, finance, institutions, and the very innovation process itself. Moreover you will recognize that these bundles of innovations must be tailored to different agri-food systems and the outcomes that need to be strengthened within them. Current agri-food systems are unsustainable in so many ways. This body of work provides many ways to turn this around—all backed up by a balanced, scrupulous use of data, evidence and reason. In a world wrongly screaming out for simple answers to complex problems, this work provides simple approaches to negotiating the complexity of context and problems to arrive at bundles of innovations that address the issues of today—and the next 30 years."

—Lawrence Haddad, Ph.D., *Executive Director, Global Alliance for Improved Nutrition, 2018 World Food Prize laureate*

"Global agri-food systems face several challenges that call for collective action at all levels. The scientific panel that authored this book employed rigorous and robust analysis to generate actionable recommendations to successfully transform agri-food systems. These recommendations can positively impact the lives and livelihoods of everyone, especially the poorest people on Earth, by advancing healthy, equitable, resilient and sustainable agri-food systems."

—Mandefro Nigussie, Ph.D., *CEO, Agricultural Transformation Agency of Ethiopia*

"This book draws on the knowledge of 23 world renowned experts to make an outstanding contribution to our understanding of food systems and how they should be changed, as we prepare for an uncertain future. A set of recommended actions is based on a comprehensive and transdisciplinary assessment of the relevant evidence and the book is truly a must-read for anybody interested in current and future food systems that benefit people and the environment."

—Per Pinstrup-Andersen, Ph.D., *Professor Emeritus, Cornell University, Adjunct Professor, Copenhagen University, 2001 World Food Prize Laureate*

"For the global agricultural research community, the research synthesis and central messages of this book are timely and vital. Innovation is multi-faceted, and is about partnership above all. New technologies and knowledge must travel hand-in-hand with facilitative policies and institutions. The authors challenge us to strive beyond our scientific work to invest in the collaborative innovation that drives meaningful and lasting transformation of our food, land and water systems."

—Claudia Sadoff, Ph.D., *Executive Management Team Convener and Managing Director, Research Delivery and Impact, CGIAR System Organization*

"Finally, this comprehensive review has put together the seven essential actions that must be included when developing "healthy, equitable, resilient and sustainable" agri-food systems! I wish I knew these ideas decades ago when our focus was on participatory testing of improved technologies in major R&D initiatives. I hope the One CGIAR takes a very careful look. Thanks to Chris Barrett and the impressive team."

—Pedro Sanchez, Ph.D., *Research Professor, University of Florida; 2002 World Food Prize Laureate*

"This is an important volume. Its core perspective is that the technological potential is huge, but that successful innovation requires enabling market, regulatory, and sociocultural environments—and that these are context specific. The report makes a strong and convincing case that bundles of context-dependent and mutually reinforcing socio-technical innovations are essential for the much-needed transition of our global food systems."

—Johan Swinnen, Ph.D., *Director General, International Food Policy Research Institute*

"The agri-food systems transformation necessary to deliver a healthy, equitable, resilient and sustainable (HERS) future must embrace context-specific innovations that can scale and deliver impacts on the 2030 Agenda for Sustainable Development and beyond. This book synthesizes a wealth of evidence, building on the scientific and engineering advances that can unlock growth in production while safeguarding our planet and improving equity and inclusion, as well as the policies, investments, governance and related social, economic and political changes necessary to incentivize and implement transformative innovations. This book is a vital contribution to the deliberations of the UN Food Systems Summit 2021, and the proposed solutions for food systems transformation."

—Shakuntala Thilsted, Ph.D. *Global Lead for Nutrition and Public Health, WorldFish; 2021 World Food Prize Laureate*

Contents

Socio-Technical Innovation Bundles for Agri-Food Systems Transformation 1

The State of Agri-Food Systems and Agri-Food Value Chains in 2020 21

Key External Drivers of Change to 2070 47

Envisioning Four Design Objectives for 2045–2070 55

Getting from Here to There 65

A Profuse Pipeline of Promising Options 73

Socio-Technical Innovation Bundles Tailored to Distinct Agri-Food Systems 159

Impact Pathways 169

Towards Co-Creation of AFS Innovations by AVC Actors 177

Technical Appendix 187

Index 191

Panel Members

Co-Chairs

Christopher B. Barrett is the Stephen B. and Janice G. Ashley Professor of Applied Economics and Management, and an International Professor of agriculture at the Charles H. Dyson School of Applied Economics and Management, as well as a professor in the Departments of Economics and of Global Development and in the Jeb E. Brooks School of Public Policy, all at Cornell University. He is also co-editor-in-chief of the journal *Food Policy*.

Tim Benton is research director in energy, environment, and resources at the Royal Institute of International Affairs at Chatham House and professor at the University of Leeds.

Jessica Fanzo is the Bloomberg Distinguished Professor of Food Policy and Ethics at Johns Hopkins University.

Mario Herrero is a professor of sustainable food systems and global change in the Department of Global Development, a Cornell Atkinson Scholar, and Nancy and Peter Meinig Family Investigator in the Life Sciences, all at Cornell University. At the time of writing, he was chief research scientist of agriculture and food at Australia' Commonwealth Scientific and Industrial Research Organisation (CSIRO).

Rebecca J. Nelson is a professor in the School of Integrative Plant Science (Plant Pathology and Plant-Microbe Biology, and Plant Breeding and Genetics Sections) and the Department of Global Development at Cornell University.

Members

Edward Buckler is a United States Department of Agriculture—Agricultural Research Service (USDA-ARS) research geneticist and adjunct professor in Plant Breeding and Genetics at Cornell University.

Karen Cooper is the R&D program manager for climate change at Nestlé, with a background in nutrition, sustainable food systems, and product innovation.

Karrie Denniston serves as senior director of sustainability with Walmart.org.

Shenggen Fan (樊胜根) is chair professor of China Agricultural University and was Director General of the International Food Policy Research Institute from 2009–2019.

Rikin Gandhi co-founded and is the executive director of Digital Green.

Steven James is a senior director of global procurement at PepsiCo.

Mark Kahn is a managing partner of Omnivore, an agritech venture capital firm based in India.

Laté Lawson-Lartego is Oxfam America's food systems theme department director.

Alexander Mathys is a professor in the Department of Health Sciences and Technology at ETH Zurich, where he leads the Sustainable Food Processing group.

Andrew Mude is the division manager of agricultural research, production, and sustainability at African Development Bank.

Felix Preston is the director of sustainability insights at Generation Investment Management.

Howard-Yana Shapiro was chief agricultural officer and Mars fellow at Mars, Inc.

Jianbo Shen is a professor in the Department of Plant Nutrition and vice dean for the National Academy of Agriculture Green Development in China Agricultural University.

Lindiwe Majele Sibanda is director of the African Research Universities Alliance (ARUA) Centre of Excellence in Food Security.

Roy Steiner is senior vice president for the Food Initiative at The Rockefeller Foundation.

Philip Thornton is a principal scientist and leads the "Priorities and Policies for Climate-Smart Agriculture" flagship of the CGIAR research program on Climate Change, Agriculture, and Food Security (CCAFS) at the International Livestock Research Institute (ILRI), Nairobi.

Stephen Wood is a senior scientist, Agriculture and Food Systems at The Nature Conservancy and associate research scientist at the Yale School of the Environment.

Special Thanks

Isatou Jallow founder and executive director of AfriCAN, the Africa Catalyzing Action for Nutrition Network, participated in the initial panel workshop in December 2019 but had to drop off the panel due to a family emergency.

Contributing Coauthors

Elizabeth Bageant is an applied research and outreach manager in the Charles H. Dyson School of Applied Economics and Management at Cornell University.

Isabella Culotta is an undergraduate research assistant at Cornell University studying plant science and international agriculture and rural development.

Jiali Liu is a graduate student at Yale University and was an undergraduate research assistant at Cornell University studying economics and statistical sciences.

Quinn Marshall is a human nutrition doctoral candidate and Center for a Livable Future–Lerner fellow at the Johns Hopkins Bloomberg School of Public Health.

Daniel Mason-D'Croz is a Research Associate in the Department of Global Development, Cornell University. At the time of writing, he was a senior research scientist at CSIRO.

Cynthia Mathys is a senior manager of strategic partnerships at the Cornell Atkinson Center for Sustainability.

Veronica Mazariegos-Anastassiou is a co-founder and farmer at Brisa de Año Ranch, a small-scale organic farm in Pescadero, CA, and recently earned a master's degree in applied economics and management from Cornell University.

Alesha (Black) Miller is the vice president of strategy and partnerships at Digital Green.

Kamakhya Misra is a corporate finance (M&A) analyst at Rabobank North America and a former undergraduate research assistant at Cornell University.

Claire Song is an undergraduate operations intern at Cornell University studying government, and environment and sustainability.

Nature Research Editors

Monica Contestabile is the chief editor at *Nature Sustainability*.

William Burnside is a senior editor at *Nature Sustainability*.

Anne Mullen is the chief editor at *Nature Food*.

Acronyms

AE	Agroecology
AfDB	African Development Bank
AFS	Agri-Food System
AMC	Advanced Market Commitment
AVC	Agri-Food Value Chain
BSF	Black Soldier Fly
CAU	China Agricultural University
CEA	Controlled Environment Agriculture
CSIRO	Commonwealth Scientific and Industrial Research Organisation
EFSA	European Food Safety Authority
ESG	Environment, Social, and Governance
EU	European Union
FAO	Food and Agriculture Organization
GDP	Gross Domestic Product
GHG	Greenhouse Gas
GM	Genetic Modification
GMO	Transgenic Crops
HERS	Healthy, Equitable, Resilient, Sustainable
HHSS	Human (H) Agency, Heterogeneity (H), Spillover (S) Effects, and Scientific (S) Research
ICMA	International Capital Market Association
ILO	International Labour Organization
IPM	Integrated Pest Management
IPSF	International Platform on Sustainable Finance
IRRI	International Rice Research Institute
ISFM	Integrated Soil Fertility Management

ISO	International Organization for Standardization
KPM	Key Performance Measure
LMIC	Low- and Middle-Income Country
MHT	Multi-Hurdle Technology
OECD	Organization for Economic Cooperation and Development
P2P	Peer-to-Peer
PES	Payments for Ecosystem Services
PN	Personalized Nutrition
PPE	Personal Protective Equipment
PRSV	Papaya Ringspot Virus
R&D	Research and Development
RAS	Recirculating Aquaculture System
SBT	Science-Based Target
SCP	Single Cell Proteins
SDG	United Nations Sustainable Development Goal
SOM	Soil Organic Matter
SSB	Sugar-Sweetened Beverages
STB	Science and Technology Backyard
TFP	Total Factor Productivity
UN	United Nations
WHO	World Health Organization

List of Figures

Socio-Technical Innovation Bundles for Agri-Food Systems Transformation

Fig. 1 The agri-food systems innovation cycle 6
Fig. 2 The 17 sustainable development goals 11

The State of Agri-Food Systems and Agri-Food Value Chains in 2020

Fig. 1 Average maize (corn) yields in the United States, 1866–2014 22
Fig. 2 Trends in agricultural land and labor productivity, 1961–2016 23
Fig. 3 Global crop yield, labor, and land productivity growth rates, 1960–2020 24
Fig. 4 Global population undernourished, 2000–2019 26
Fig. 5 Global real food prices, January 1990–July 2020 27
Fig. 6 Shifting food consumption patterns with income growth 28

Key External Drivers of Change to 2070

Fig. 1 Human population projections by world region, 1950–2100 48
Fig. 2 Off-farm share of agri-food system GDP and employment by income per capita 50

Envisioning Four Design Objectives for 2045–2070

Fig. 1 Long-term cereal yields of key crops in the United Kingdom from 1270 to 2018 57
Fig. 2 Global crop yields from 1961 to 2018 58

Getting from Here to There

Fig. 1 Stunting and obesity by system type 66
Fig. 2 The nutrition transition in five patterns 67
Fig. 3 Share of food purchases by type of vendor 68
Fig. 4 Online grocery sales trends, 2012–2019 68

A Profuse Pipeline of Promising Options

Fig. 1 Promising emergent technologies span the AVC 74
Fig. 2 Technological readiness of future agri-food systems technologies 76
Fig. 3 The digital agri-stack 79
Fig. 4 Social protection program coverage among 108 low- and middle-income countries 94
Fig. 5 How different social protection measures fit together 96

Socio-Technical Innovation Bundles Tailored to Distinct Agri-Food Systems

Fig. 1 Essential elements for accelerating the systemic transformation of agri-food systems 162
Fig. 2 Socio-technical bundles fit for purpose to an objective and context 167

Impact Pathways

Fig. 1 Net impacts of different technology domains on food systems–related SDGs and their indirect effects on other SDGs 170
Fig. 2 Range of potential direct impacts of anticipated technologies across SDGs 171
Fig. 3 Potential impact pathways of personalized nutrition and nitrogen-fixing cereals 173

Towards Co-Creation of AFS Innovations by AVC Actors

Fig. 1 Coordinated-action agenda regarding food product standards ... 183

List of Tables

Socio-Technical Innovation Bundles for Agri-Food Systems Transformation

Table 1 Human population and land area by agri-food system type 15

Socio-Technical Innovation Bundles Tailored to Distinct Agri-Food Systems

Table 1 Accelerators for two promising agri-food technologies 163

List of Boxes

Key External Drivers of Change to 2070

Box 1 Turn attention to Africa 49

A Profuse Pipeline of Promising Options

Box 1 Prioritizing interventions for climate-smart agri-food systems 77
Box 2 Transgenic and gene editing technologies 102
Box 3 Science and technology backyards—linking farmers, extension, agribusiness, and science at scale 106
Box 4 Regulatory nudges towards integrated pest management 112
Box 5 Microbial, insect, and algal biomass as circular feeds 118
Box 6 Towards fact-based sustainability labelling 126
Box 7 Reformulation, fortification, and functionalization—incentivizing old innovations 135

Socio-Technical Innovation Bundles for Agri-Food Systems Transformation

> *Innovation, like evolution, is a process of constantly discovering ways of rearranging the world into forms that are unlikely to arise by chance—and that happen to be useful . . . . [I]nnovation is the most important fact about the modern world, but one of the least well understood. . . . The striking thing about innovation is how mysterious it still is. No [scientist] can fully explain why innovation happens, let alone why it happens when and where it does.*
>
> —Matt Ridley, *How Innovation Works* (2020)

Technological and institutional innovations in agri-food systems (AFSs)[1] over the past century have brought dramatic advances in human well-being worldwide. Yet these gains increasingly appear unsustainable due to massive, adverse spillover effects on climate, natural environment, public health and nutrition, and social justice (Barrett 2021). How can humanity innovate further to bring about AFS transformations that can sustain and expand past progress, while making them healthier for all people and for the planet that must sustain current and future generations?

[1] We favor the "agri-food" modifier of "systems" and "value chains" because the value chain transforms the agricultural feedstocks produced by farms, fisheries, and natural harvest into the foods humans eat. Many farms and fisheries cultivate both food and non-food products (e.g., cotton; sisal; tobacco; or fish glue, meal, or oil). And people consume little food that has not been packaged, prepared, processed, or transported off-farm/fishery. Therefore, both the "agricultural" and "food" modifiers are too narrow on their own. Note that we include both wild capture and domesticated production of animals and plants of all sorts under the "agri-food" label.

C. B. Barrett et al., *Socio-Technical Innovation Bundles for Agri-Food Systems Transformation*, Sustainable Development Goals Series,
https://doi.org/10.1007/978-3-030-88802-2_1

Recent scientific studies of global AFSs bring out clearly the challenges we face. Some emphasize the environmental and climate unsustainability of AFSs (GloPan 2016, 2020; IPCC 2019; IPBES 2019; Willett et al. 2019). Given projected growth in human populations and incomes, and the headwinds of the climate and extinction crises, satisfying future aggregate demand for food will put unprecedented pressures on finite water, land, genetic, and atmospheric resources. The risks of enormous and potentially irreversible ecological damage are no longer under serious scholarly dispute. Moreover, beyond the longer-run pressures wrought by inevitable food-demand growth, building evidence raises concerns about AFSs' resilience to sudden weather, environmental, disease, economic, or political shocks. Such shocks appear to be rising in frequency and/or intensity, and commonly cascade, with one triggering another (Maystadt and Ecker 2014; Von Uexkull et al. 2016). And shocks to AFSs increasingly appear to feed sociopolitical instability around the world in a potentially vicious cycle (Barrett 2013).

Other recent studies point to AFSs' failure to advance the well-being of all persons, in at least two distinct ways (GloPan 2016, 2020; Haddad et al. 2016; FAO 2019; HLPE 2020). First, today's AFSs fail to ensure healthy diets for all—a necessary condition for food security.[2] Second, AFSs do not provide equitable and inclusive livelihoods for the roughly half of the world's labor force—more than 1.3 billion people (ILO 2015)—who work in agri-food value chains (AVCs).[3] Far too many people who labor on farms, in processing facilities, groceries, restaurants, or elsewhere within our AFSs fail to earn a living income or to control essential resources such as land, or risk serious injury or illness, or are victims of forced labor. Women, indigenous populations, racial and religious minorities, and young people are disproportionately disadvantaged for a variety

[2] We rely on the definition agreed to by all parties to the 1996 World Food Summit: "Food security exists when all people, at all times, have physical and economic access to sufficient, safe and nutritious food that meets their dietary needs and food preferences for an active and healthy life."

[3] AVCs encompass pre-farmgate input suppliers as well as the whole post-farmgate range of processing, storage, transport, wholesaling, retailing, food service, and other functions that transform the agricultural outputs that farms, fisheries, and natural harvesters produce into the foods humans consume multiple times every day. Relative to food systems, the AVC focuses attention on human agency, on the myriad actors whose choices individually and collectively drive food-systems evolution. Desirable systems change requires human behavioral change, hence our focus on AVCs so as to emphasize human agency.

of systemic reasons. Despite the unprecedented productivity and prosperity enabled by technological advances and institutional and policy reforms in global AFSs over the past century, far too many people still face chronic or episodic undernutrition, diet-related health risks are a growing problem, and AFS jobs are among the most dangerous and exploitation-prone on the planet.

The COVID-19 pandemic has laid bare previously under-recognized fragilities that pose yet another hidden cost of the modern AFS: uninsured risk of catastrophic disruptions. Past, sometimes-single-minded pursuit of lower food production costs and consumer prices brought valuable efficiency gains. But it has also led to such AVC specialization and concentration based on economies of scale and scope that many producers and sub-systems struggled to adjust to a massive systemic shock. Advances in logistics and market integration enabled reasonably quick stabilization of food supplies and prices in most places. But we should understand the COVID-19 pandemic as a warning shot across the bow of AFSs. As scientists expect natural and manmade shocks to grow in frequency and severity, enhancing AFS resilience grows ever more urgent and may entail building in some redundancy as systemic insurance (Webb et al. 2020).

This book originated as a report commissioned by the Cornell Atkinson Center for Sustainability in response to an invitation from the journal *Nature Sustainability*, which—in collaboration with its new sister journal, *Nature Food*—wanted to devote its 2020 expert panel to this topic.[4] The panel brought together experts who come from many different continents and who span a wide range of disciplines and organizations—from industry and universities to social movements, governments, philanthropies, institutional and venture capital investors, and multilateral agencies.

The panel synthesized the best current science to describe the present state of the world's AFSs and key external drivers of AFS changes over the next 25–50 years, as well as tease out key lessons from the COVID-19 pandemic experience this year. As is increasingly widely recognized, **the costs that farmers and downstream value chain actors incur and the prices consumers**

AS IS INCREASINGLY WIDELY RECOGNIZED, THE COSTS THAT FARMERS AND DOWNSTREAM VALUE CHAIN ACTORS INCUR AND THE PRICES CONSUMERS PAY UNDERSTATE FOODS' TRUE COSTS TO SOCIETY ONCE ONE ACCOUNTS FOR ADVERSE ENVIRONMENTAL, HEALTH, AND SOCIAL SPILLOVER EFFECTS.

[4] *Nature Sustainability* endorses one expert panel per year. The first, on science and the future of cities, convened in 2018, and the second, on behavioral science for design, in 2019.

pay understate foods' true costs to society once one accounts for adverse environmental, health, and social spillover effects. Inevitable demographic, economic, and climate change in the coming decades will catastrophically aggravate these problems under business-as-usual scenarios. Innovations will be needed to facilitate concerted, coordinated efforts to transition to more healthy, equitable, resilient, and sustainable AFSs.

In deliberating about needed innovations, the panel concluded that four key AFS features must continuously remain front-of-mind: decentralized individual and collective **h**uman (H) agency that drives systemic change, the intrinsic **h**eterogeneity (H) of AFSs locally and globally, pervasive **s**pillover (S) effects, and the essential role of **s**cientific (S) research. Attention to these HHSS (pronounced "his") attributes is essential to avoid adverse unintended consequences and make real progress.

The panel then developed a shared vision for the AFSs of 2045–70, beyond the 2030 horizon of the UN Sustainable Development Goals (SDGs). We summarize that vision in four core AFS objectives: **h**ealthy (H) and nutritious diets, **e**quitable (E) and inclusive value chains, **r**esilience (R) to shocks and stressors, and climate and environmental **s**ustainability (S), summarized in the acronym HERS. **AFSs are immutably HHSS. The task is to make them equally HERS.** Failure to address the HERS objectives risks catastrophic failure, even existential threats, under business-as-usual scenarios. Faced with multiple, high-level, pressing objectives, AFS adaptations cannot attend only to unidimensional concerns, whether about climate, environment, health, employment, equity, productivity, or resilience. Both tradeoffs and synergies exist among these design objectives. For that reason, among others, we therefore need bundled responses to address looming challenges and to realize the considerable promise of a rich pipeline of emergent technologies, a portfolio to deliver on multiple objectives that no one innovation can simultaneously satisfy.

With a shared assessment of current state—and of inexorable drivers of AFS change—and a shared vision of desired future state firmly in mind, the panel then undertook a detailed review of scores of innovations at various stages of development and implementation.[5] The pipeline of emergent technologies is full of

[5] Because we are looking into the future, in some cases by decades, little if any rigorous impact evaluation evidence exists on the innovations we discuss. We rely to the maximum extent possible on limited model-based, carefully reasoned, or suggestive empirical evidence that exists, and we cite those sources for readers. Innovations necessarily require rigorous monitoring and evaluation as they diffuse and scale, so as to ensure wise management of scarce natural, human, and financial resources.

promise.[6] A disproportionate share of them are digital innovations, but the abundance of agronomic, genetic, mechanical, and social science advances available to advance HERS objectives is undeniable. One cannot help but conclude that existing and imminent knowledge really are not the factors limiting progress in addressing the formidable challenges facing AFSs now and in the coming years.

The limiting factor is more sociopolitical: insufficient leadership, political will, and willingness to find cooperative solutions rather than winner-take-all outcomes. All new technologies must navigate a complex maze of biophysical, political economy, and sociocultural obstacles to adapt and scale, and thus they need companion interventions to accelerate them to implementation and diffusion. Furthermore, every innovation we studied will almost surely have unintended impacts on non-target outcomes, and the resulting tradeoffs naturally spark opposition by groups concerned that change might hurt them. The panel therefore heavily emphasized the importance of coupling technical advances with social and policy change, into socio-technical innovation bundles customized to each AFS context's needs to realize the HERS objectives. But **identifying and bundling the right innovations is an intrinsically social process, one that demands cooperation that is in shorter supply than are brilliant scientific insights.**

This can be summarized in the conceptualization of the AFS innovation cycle depicted in Fig. 1. Human agency drives the AFS innovation cycle. External drivers (e.g., demographic change, income growth, climate change) influence collective objectives (e.g., HERS outcomes) and actor-specific objectives (e.g., firm profits or political power) and, jointly with those objectives, induce myriad innovations by individuals

IDENTIFYING AND BUNDLING THE RIGHT INNOVATIONS IS AN INTRINSICALLY SOCIAL PROCESS, ONE THAT DEMANDS COOPERATION THAT IS IN SHORTER SUPPLY THAN ARE BRILLIANT SCIENTIFIC INSIGHTS.

[6]An online collaborative web portal is expected to launch in early 2021, hosted as a sub-domain of the NutritionConnect (https://nutritionconnect.org/) site. This is a joint effort between our expert panel; the CSIRO Wild Futures Project (Herrero et al. 2020, 2021); and Project Disrupt: Healthy Diets on a Healthy Planet, a three-stage Delphi study jointly led by the Global Alliance for Improved Nutrition, the Alliance of Bioversity International and the International Center for Tropical Agriculture, and EAT. The aim of the portal is to facilitate discovery and contribution of information on food systems innovations, of prospective collaborators, and of opportunities for cross-system and cross-sector learning.

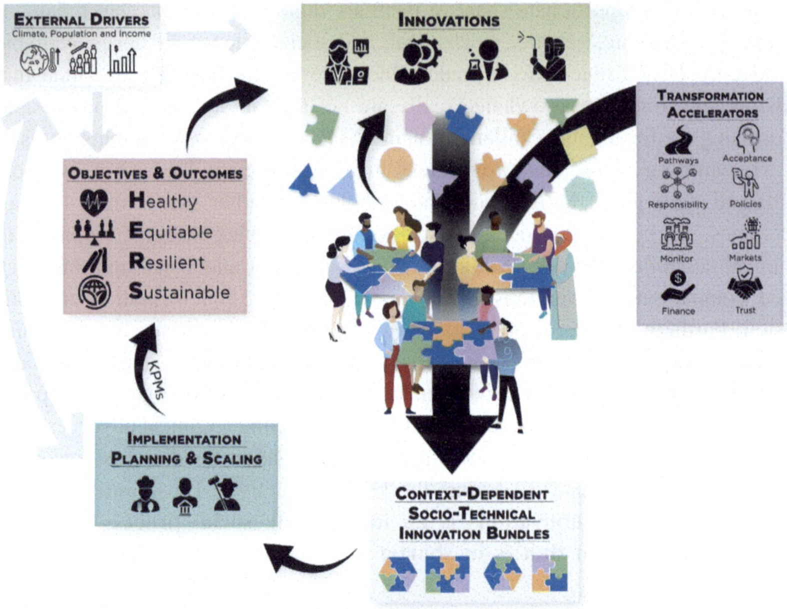

Fig. 1 The agri-food systems innovation cycle

and organizations. Innovations (represented by puzzle pieces) draw on different (natural or social) science-based methods (represented by different colors) to generate products, processes, or policies with distinct designs and purposes (represented by different shapes). Transformation accelerators—key enabling societal features—help AFS-specific stakeholders redirect some ill-fitting innovations back for adaptation to the local context and accelerate combination of other innovations. To become implementable and scalable, socio-technical innovation bundles need appropriate, context-dependent pieces and the right composite shape to fit local purposes. Implementation and scaling then generate feedback that affects external drivers, and in combination with those external drivers, generate outcomes. Monitoring key performance measures (KPMs) informs assessment of those outcomes—and of individual and combinatorial innovations—and helps direct adaptive management of synergies and tradeoffs among objectives, renewing the AFS innovation cycle.

Co-creation of socio-technical innovation bundles necessarily requires multi-party cooperation among public and private sector organizations. The panel therefore developed some process and action recommendations to guide AVC actors as we navigate together from the present, precarious state to a HERS one in our children's future. This requires some basic rules of engagement, including discussing KPMs to monitor progress. After all, we manage what we measure. Significant public investment and trust in first-rate science will be necessary but far from sufficient. Investment increasingly turns on performance assessed, for better or worse, by KPMs. The institutional, policy, and sociocultural accelerators of technological adaptation, diffusion, and upscaling are essential complements. Hence the need for different AVC actors' active engagement in the AFS innovation cycle.

One central message of the report is that in championing the foundational role of science and engineering to enable sustainable progress, too many high-level reports inadvertently downplay the equally crucial role of human agency (NASEM 2020). We therefore focus not only on prospective innovations but just as much on the necessary actions by actors throughout AVCs.

Change only comes about through the actions of people and the organizations they comprise. Impactful innovation can originate among actors anywhere along the food value chain, induced by any of a host of motives. So, too, can obstruction. Throughout human history, the greatest progress has come through innovation, be it in biophysiochemical technologies (e.g., improved plant and animal genetics; new medicines, transport, or computing equipment) or institutions[7] (e.g., formal policies such as rules of tenure over land and water or contract law, or informal sociocultural practices such as cuisine). In order to harness the potential of the breathtaking pace of innovation today in digital, genetic, and other spaces, many different actors—consumers, retailers, restaurants, distributors, processors, farmers, input manufacturers, governments, charitable organizations, etc.—must engage in honest, constructive dialogues of the sort we undertook with the objective of co-designing contextually appropriate socio-technical bundles of innovations that can enable navigation away from looming dangers and towards a HERS future.

[7]We use the definition promulgated by the Nobel Laureate Douglass C. North (1991): "Institutions are the humanly devised constraints that structure political, economic, and social interaction. They consist of both informal constraints (sanctions, taboos, customs, traditions, and codes of conduct), and formal rules (constitutions, laws, property rights)."

In order to enjoy HERS agri-food systems at a horizon of 25–50 years, we must invest and innovate today. We will reap then what we sow now. Innovation takes time. The lag from scientific discovery to its implementation in new technologies to productivity or other improvements at sufficient scale to be detectable in industrial, sectoral or national data is typically 15–25 years (Adams 1990; Chavas et al. 1997; Ahmadpoor and Jones 2017; Baldos et al. 2019).[8] This compels decentralized, coordinated action by public, private, and civil society actors throughout AVCs, starting immediately. Redirecting the course of AFSs presently headed towards climate, environmental, public health, and social justice disaster will require all hands on deck, working together with shared responsibility to do the hard work of navigating away from danger and towards environmentally and socially sustainable AFSs to sustain future generations. We are concerned, but ultimately optimistic that from the grim turmoil of 2020 will emerge greater unity and resolve to successfully address the systemic issues that bedevil AFSs locally and globally and that imperil our children's and grandchildren's futures.

Ultimately, the analysis presented in the ensuing pages culminates in **seven essential actions** that must guide agri-food systems transformations. In no particular order, these are:

Develop socio-technical innovation bundles: Despite the abundance of rapidly progressing innovations across all stages of AVCs today—in digital, genetic, and other spaces—**no magic scientific or engineering bullets exist**. Few, if any, innovations can adapt and scale effectively without essential supporting policies and institutions. Innovation is as much a social process as a scientific one, and no innovation we could identify can effectively target all four HERS objectives simultaneously. We therefore need a portfolio approach to deliver impact and to maintain necessary balance among objectives. The creative destruction of technological change inevitably generates both winners and losers, and new technologies will almost surely produce both positive and negative spillovers across HERS objectives. Co-creation of bundled approaches is therefore essential to enable packages of new technologies and practices to emerge, adapt, and diffuse to scale within, and across, contexts, and to generate beneficial impacts with limited, or no unintended, net adverse consequences.

[8] The estimated lags vary by the discipline of discovery, with more basic sciences like mathematics generating impact with longer lags than more applied ones, such as computer science (Ahmadpoor and Jones 2017) and private R&D investments generating larger near-term–in the 5–15 year window–payoffs, with public R&D delivering bigger longer-term gains at 15–25 year horizons (Chavas et al. 1997).

Reduce the land and water footprint of food: Meeting future growth in food demand while reducing AFS land and water use is both necessary and inevitable. We cannot effectively tackle the climate and extinction crises and reduce the risk of zoonosis-driven pandemics without reducing AFS terrestrial and marine footprints. **Decoupling food demand growth from land and water use is perhaps the most essential and challenging transition task we face**. That process must be actively and cooperatively negotiated among diverse stakeholders.

Commit to co-creation with shared and verifiable responsibility: The complex pathways from innovation to scaling to impact necessitate co-creation of locally contextualized socio-technical bundles. Because human agency drives everything, all parties need incentives to act, including explicit sharing of both the responsibility to address emergent challenges and the benefits from innovation. Shared responsibilities must be matched with verifiable key performance metrics, agreed sanctions for transgressions, and safety-net protections against losses. Co-designed socio-technical bundles can accelerate human agency to facilitate, rather than obstruct, beneficial innovation and minimize unintended consequences.

Deconcentrate power: Many components of candidate solutions are well known, but impeded by concentrated economic and political power or by the marginalization of key stakeholders. The powerful can too easily obstruct progress (e.g., via catch-and-kill acquisitions, political lobbying, patent thickets). Reducing market and political power imbalances and broadening participation in innovation dialogues can accelerate innovation. Novel financing of discovery for open-source innovation, reforms of intellectual property regimes, and more robust enforcement of anti-trust laws can accelerate beneficial transitions, as can more concerted government and civil society efforts to facilitate participatory dialogues to foster co-creation of effective solutions.

Mainstream systemic risk management: The COVID-19 pandemic underscores the rising importance of building effective systemic risk management for AFSs. Most governments already appropriately mandate many forms of individual insurance (automobile, fire, health, etc.) so as to resolve market failures and avert catastrophic spillover effects. We increasingly need analogous approaches—both risk reduction and risk transfer mechanisms—to address low-probability, high-impact events (e.g., pandemics) or a combination of events (each with higher individual probabilities) that jointly cause a high-impact event (e.g., the 2007–08 food price spike).[9]

Develop novel financing mechanisms: AFS innovations and systemic risk management require massive up-front investment of hundreds of billions of dollars additional resources annually. This is feasible but demands creativity, especially to mobilize private resources beyond public spending and philanthropic investments. The world is awash in investible resources, with historically low interest rates and high equity market valuations. The COVID-19 pandemic has proved that governments can quickly mobilize massive public funding when the stakes are high and solutions are urgently

[9] See, for example, Barrett (2013), Homer-Dixon et al. (2015), and Challinor et al. (2018).

needed. Meanwhile, a growing community of private investors recognize the complementarities between longer-term financial and non-financial outcomes. Novel methods to mobilize the financing necessary for transforming AFSs are rapidly emerging.

Reconfigure public support for AFSs: Governments play two essential roles: investing in essential public goods and services—including basic science and education, reliable data, and appropriate, effective regulation—and facilitating dialogue to find cooperative solutions. Far too much current government agri-food spending is misspent, especially the roughly US$2 billion/day that goes to environmentally harmful farm subsidies that impede necessary innovation and disproportionately benefit better-off landowners, many of whom do not actively farm themselves. Governments must crowd in far greater private investment in AFS transformations by redirecting public resources towards social protection programs, agri-food research, and physical and institutional infrastructure (e.g., universal rural broadband access, extension services, product standards, food safety assurance). Governments also play an essential role convening civil society dialogues to facilitate discovery of, and support for, appropriate socio-technical bundles. Governments likewise must lead in co-developing and endorsing commitment frameworks, and complementary indicators and accountability mechanisms to ensure effective implementation of identified cooperative solutions at national, regional, and global scales.

Four Key Features of Agri-Food Systems and Agri-Food Value Chains

As the first-ever United Nations (UN) Food Systems summit approaches in 2021, many people and organizations are thinking carefully about how to transform contemporary AFSs to more effectively advance the 17 SDGs (Fig. 2) set in 2015 by the UN General Assembly with the intention of achieving each of them by 2030. SDG 2 (Zero Hunger) perhaps draws most attention in discussions of AFSs, but strong connections exist to virtually every one of the other 16 SDGs as well.

AFSs consist of webs of interactions among human actors, non-human organisms, and abiotic processes, with complex interlinkages across trophic scales, economic sectors, geographic space, and time. **Everything that goes into growing, capturing, storing, transforming, distributing, or eating food fits within AFSs**. The literature is rich with various representations of AFSs (Ericksen 2008; Global Panel on Agriculture and Food Systems for Nutrition 2016; HLPE 2017; Fanzo et al. 2020), all of which necessarily oversimplify so as to emphasize specific foci appropriate to their immediate purpose. But across the myriad AFS

Fig. 2 The 17 sustainable development goals

depictions, the four key HHSS features stand out as especially relevant when trying to promote beneficial innovations.

Inevitably decentralized decision-making within AFSs underscores the first key feature of agri-food systems: **human agency.** Our emphasis on AVCs follows from the centrality of decentralized exercise of human agency by actors each pursuing objectives that may, at times, conflict with one another. Command-and-control systems do not work because the interests of the powerful still prevail, even if power is conferred through political rather than market processes. Rather, societies must find ways to reconcile multiple, sometimes-competing objectives in pluralistic systems.

EVERYTHING THAT GOES INTO GROWING, CAPTURING, STORING, TRANSFORMING, DISTRIBUTING, OR EATING FOOD FITS WITHIN AGRI-FOOD SYSTEMS.

This often means fostering collective action. Hence the importance of mechanisms to improve coordination and align incentives, and the generation of behaviors that produce positive externalities, as well as of innovations to reduce negative externalities in those areas where coordination routinely fails. The structures and processes through which people and organizations acquire, maintain, and exercise sociopolitical power and cultural influence matter enormously to whether, and what sorts of, coordination will emerge. Hence the rising global chorus for more explicitly incorporating human agency in the conceptualization of food security, so as to elevate the right to food already recognized in treaties, including Article 25 of the 1948 Universal Declaration of Human Rights; Article 11.2 of the 1966 International Covenant on Economic, Social, and Cultural Rights; and in the constitutions of dozens of countries (Vidar et al. 2014; Gundersen 2019; HLPE 2020).

In recognizing the central role of human agency in AFSs, we also need to avoid the common temptation to focus excessively on either end of the value chain: upstream farmers and/or downstream food consumers. Most value addition, employment, etc., occurs between the farm and final consumer, and the relative importance of the post-farmgate stages of value chains inevitably expands with income growth and urbanization. Mid-stream value chain actors—many of them large, private corporations—too often lurk in the shadows of policy debates. These actors can, and must, be mobilized as equal partners in the co-creation of innovations to accelerate AFS transformation.

The intrinsic **heterogeneity** of AFSs is their second key feature. The coordination mechanisms and science necessary to internalize or mitigate externalities so as to avoid catastrophe and to foster continuous improvement vary enormously across geographies and agroecological and socioeconomic contexts. **One-size-fits-all solutions do not exist**. The panel therefore eschews ranking specific innovations, as performance will typically vary by context.

We adopt the approach of the Food Systems Dashboard, an excellent new tool that curates myriad data sources to enable visualization of key data series at country, regional, and global scales, and emphasizes five AFS types (Fanzo et al. 2020; Marshall et al. 2021)[10]:

1. **Rural and traditional**: Farming is dominated by smallholders, and agricultural yields are typically low. Most farmers focus on staple crops (and retain much of their harvest for their own consumption) and a limited number of cash crops. Food imports and exports represent a small percentage of domestic consumption and production. Supply chains are short, resulting in many local, fragmented markets and limited non-farm AFS employment. Limited cold chains and storage facilities cause large food losses, which may also disincentivize diversification into perishable foods. The quantity and diversity of foods available varies seasonally, often with a pronounced lean season. Food is mainly sold through informal market outlets, including independently owned

[10] See the Technical Appendix for further detail, drawing on Marshall et al. (2021), which details the methodology underpinning the identification of these food system typologies.

small shops, street vendors, and periodic markets. Supermarkets are uncommon, especially outside of major cities. Mandatory or voluntary fortification guidelines for staple foods are common in order to combat micronutrient deficiencies.

2. **Informal and expanding**: Average agricultural land and labor productivity and access to inputs (e.g., improved seeds and fertilizer) are higher than in traditional systems and rising. Modern food supply chains are in place for grains and other dry foods, which include processors and centralized distribution centers. These are also emerging for fresh foods, though traditional supply chains continue to dominate due to cold chains and other market infrastructure that remain underdeveloped. Processed and packaged foods are available in both urban and rural areas. Food processing may incorporate a combination of locally sourced and imported ingredients. Demand for convenience foods increases as the formal, non-farm labor force grows and includes more women, with urbanization and income growth also playing a role. Supermarkets and fast food are rapidly expanding and attracting more middle-class consumers, although informal market outlets still dominate food retailing, especially for animal-source foods, fruits, and vegetables. Few food quality standards are in place and advertising is not regulated, though many countries have fortification guidelines for staple foods.
3. **Emerging and diversifying**: Large-scale commercial farms increasingly co-exist alongside large numbers of small-scale farms, all of which enjoy enhanced market integration through better communications and transportation infrastructure. Food supply chains for fresh foods, including fruits, vegetables, and animal-source foods, are developing rapidly. Supply chains are elongating, with urban areas relying on food imports and rural areas relying more on export markets than in more traditional and informal food systems. Processed and packaged foods are widely available in rural areas, with less seasonal fluctuation in availability of perishable foods. Supermarkets are common even in smaller cities, although most fresh foods continue to be purchased through informal markets. Food safety and quality standards exist, but mainly within formal markets due to limited government monitoring capacity. A greater proportion of countries in this food system type have adopted food-based dietary guidelines.
4. **Modernizing and formalizing**: Larger farms rely more on mechanization and input-intensive practices, resulting in higher agricultural land and labor productivity. Food supply chain infrastructure is more developed, resulting in fewer food losses beyond the farmgate, although waste and spoilage at the retail and consumer end of the supply chain remains a challenge. Food and

beverage manufacturing, food retailing, and food service capture a significantly greater share of consumer food expenditures. Dietary diversity rises, with regional specialization in agricultural production and imports of foods enabling more year-round availability of diverse foods. Multiple supermarket and food service chains exist within cities and larger-sized towns. These chains capture a large market share of fresh foods and are more accessible to lower-income consumers. Government regulation and monitoring of food quality standards are more common.

5. **Industrialized and consolidated**: Farming represents a land- and capital-intensive business, dominated by a small number of large-scale, input-intensive farms serving specialized domestic and international markets (e.g., horticulture, animal feed, processed food ingredients, biofuels). Market consolidation is common both upstream and downstream, as a shrinking number of large life-sciences firms supply patent-protected farm inputs while large processors, manufacturers, and retailers procure directly from farmers, reducing the number of intermediaries along the supply chain. Supermarket density is high in urban and metropolitan areas and even most medium-sized towns have access to multiple chains. The formal food sector represents nearly all domestic food consumption, including fresh foods. Luxury-oriented food retail and food service expand, creating greater quality differentiation in the food retail and food service sectors. Pockets of food insecurity still exist, often referred to as "food deserts," alongside employment, income, and wealth disparities. A greater proportion of countries in this food system type have adopted policies that ban use of industrial trans fats and reformulate processed foods for reduction of salt intake.

At a coarse scale, simply using the typology method to assign entire countries to individual AFS types drives home several key points. First, **a plurality of humanity currently lives in countries dominated by rural and traditional systems** (Table 1). Population growth and migration patterns will only reinforce this need to invest far more effort and resources in AFS innovation for the Global South. Second, most of the Earth's land mass is in the most advanced (industrial and consolidated or modernizing and formalizing) systems (Table 1). These places present especially large opportunities to transition working lands from growing food to sequestering carbon to reduce harmful greenhouse gases (GHGs) and to reap the resulting mitigation benefits: harvesting renewable energy and restoring habitats. Third, although discussions of AFSs commonly revolve around the extremes of this continuum—focusing on either the smallholder farmers that predominate in rural and traditional systems, or on the large-scale industrial farm-

Table 1 Human population and land area by agri-food system **type**

Food system type	% of global population (%)	% of global land area (%)
Rural and Traditional	31	13
Informal and Expanding	18	12
Emerging and Diversifying	24	17
Modernizing and Formalizing	11	28
Industrial and Consolidated	13	26

Source Marshall et al. (2021)

ing and food corporations of the industrial and consolidated systems—**most of the world's population resides in countries dominated by transitional states. The opportunity to shape those transitions is especially profound.**

Of course, many AFS types can co-exist within a country or even a metropolitan region. Typologies allow for cross comparisons of trends and emerging patterns at whatever level of aggregation or disaggregation the data permit. As we highlight below, the impact pathways one envisions for different innovations fundamentally turn on the characteristics of the local AFS one targets. We depict key patterns in AFSs today with reference to these five typologies.

The third key feature that stands out as especially relevant when trying to promote beneficial innovations is that the closely coupled nature of AFSs implies that actions anywhere have **spillover effects or externalities** elsewhere in the system. Examples of negative externalities abound in AFSs. Some food processing practices that reduce costs, thereby making food more affordable (e.g., by adding inexpensive fats, salt, and sugars) have adverse public health consequences. Fertilizer misuse or overuse on farms can lead to nutrient runoff into waterways that causes downstream eutrophication or harmful algae blooms that harm fisheries. Many food system processes contribute massive amounts of GHGs that adversely affect the global climate, including land clearing; tilling of soil; agrochemical applications; the digestive processes of vast numbers of ruminant livestock; and the burning of fossil fuels, either directly by farm machinery and transport equipment or indirectly by utilities that provide electricity to milking parlors, manufacturing facilities, retail outlets, etc. Equally important, however, are positive externalities that arise from other behaviors—from animal and plant disease controls that limit the spread of harmful organisms, to scientific discoveries that cascade into further innovations.

AFSs' pervasive externalities imply a divergence between the market price of foods and their social cost, once one factors in environmental, public health, and other externalities. This divergence reflects a market failure; markets typically cannot internalize spillovers easily. A range of groups are working on true cost accounting for food, often relying on life-cycle costing and similar methods to try to capture the full impacts of each product, inclusive of indirect impacts on the natural environment, public health, etc.[11] Governments must play a role in addressing the gap between market prices and true costs through regulatory, subsidy, and tax policies. But private companies and investors can do so, as well, including through innovative financing mechanisms of the sort we discuss later.

But no matter the policy instrument or pricing method governments use, they quickly confront the "food price dilemma" (Timmer et al. 1983), wherein price changes invariably cause both winners and losers. For example, higher food prices to reduce the environmental impacts of agri-food production generate environmental gains but also equity losses as foods become less affordable to the poor. Hence the central importance of technological advances—and especially socio-technical bundles—because these offer the chance to obviate the food-price dilemma and generate gains in one or more dimensions without having to impose losses on others. Advances will not always be "win–win"; a "win-neutral" is still an unambiguous improvement. The central task of innovation systems and the design of transition pathways is to identify bundles of technological and policy/social innovations that together enable what economists term "Pareto improvements" (i.e., advances for at least some without making anyone worse off).

AGRI-FOOD SYSTEMS' PERVASIVE EXTERNALITIES IMPLY A DIVERGENCE BETWEEN THE MARKET PRICE OF FOODS AND THEIR SOCIAL COST, ONCE ONE FACTORS IN ENVIRONMENTAL, PUBLIC HEALTH, AND OTHER EXTERNALITIES.

The pervasive externalities that arise from AFSs' deep connectivity through various abiotic, ecological, and human processes often induce a tempting conceit that one can optimize AFSs. But billions of individual food consumers, farmers,

[11] Examples include the Global Alliance for the Future of Food (https://futureoffood.org/impact-areas/true-cost-accounting/), and The Economics of Ecosystems and Biodiversity (TEEB 2018).

firm managers, workers, etc., make decisions and act every day, pursuing their own motives within the constraints specific to their time and station. No one has authority or control over even significant sub-systems, much less the whole. Rather, AFSs are highly decentralized networks of agents making interdependent decisions semi-autonomously. Moreover, we often overstate how well we can quantify and compare trade-offs of often fundamentally incomparable multiple objectives.

The fourth key feature of modern AFSs is the central place of **science**—for discovery, invention, adaptation, and engineering—which is necessary to maintain and advance innovation and systems performance in virtually any dimension. **The panel is alarmed by how widely—and perhaps increasingly—sound scientific advice and evidence is being ignored** by business, community, media, and political leaders, as well as by everyday decision-takers. Scientific research remains essential to unlock better ways of more efficiently using the Earth's finite resources, of combatting changing threats, and of seizing emergent opportunities. The evolutionary nature of the AFS structure implies a never-ending need for scientific research to continuously adapt to evolving systems. Hence the importance of ongoing, generous public and philanthropic funding of basic science, a pure public good on which private investors can build. Indeed, scientific discovery generates some of the greatest positive spillovers as new findings diffuse and adapt broadly throughout AFS, lowering food prices to provide consumers with more affordable and safer foods, and farmers and firms with more productive digital, genetic and mechanical inputs and management processes. **The world has previously faced daunting AFS challenges and, through science, emerged stronger; we can do it again** (Barrett 2021).

THE PANEL IS ALARMED BY HOW WIDELY—AND PERHAPS INCREASINGLY—SOUND SCIENTIFIC ADVICE AND EVIDENCE IS BEING IGNORED BY BUSINESS, COMMUNITY, MEDIA, AND POLITICAL LEADERS, AS WELL AS BY EVERYDAY DECISION-TAKERS.

Together, these four essential features of AFSs—summarized earlier in the simple mnemonic **HHSS**—must remain front-of-mind in promoting innovations within AVCs: decentralized human (H) agency, the intrinsic heterogeneity (H) of AFSs locally and globally, pervasive spillover (S) effects, and the essential role of scientific (S) research. They are foundational to the panel's assessment of the rich pipeline of emergent AFS innovations and our recommended action plans to facilitate necessary transitions in the decades ahead.

References

Adams, James D. 1990. Fundamental stocks of knowledge and productivity growth. *Journal of Political Economy* 98 (4): 673–702.

Ahmadpoor, Mohammad, and Benjamin F. Jones. 2017. The dual frontier: Patented inventions and prior scientific advance. *Science* 357 (6351): 583–587.

Baldos, Uris Lantz C., Frederi G. Viens, Thomas W. Hertel, and Keith O. Fuglie. 2019. R&D spending, knowledge capital, and agricultural productivity growth: A Bayesian approach. *American Journal of Agricultural Economics* 101 (1): 291–310.

Barrett, Christopher B (ed.). 2013. *Food security and sociopolitical stability*. Oxford: Oxford University Press.

Barrett, Christopher B. 2021. Overcoming global food security challenges through science and solidarity. *American Journal of Agricultural Economics* 103 (2): 422–447.

Challinor, Andy, W. Neil Adger, Tim G. Benton, Declan Conway, Manoj Joshi, and Dave Frame. 2018. Transmission of climate risks across sectors and borders. *Philosophical Transactions of the Royal Society a: Mathematical, Physical and Engineering Sciences* 376: 2121. https://doi.org/10.1098/rsta.2017.0301.

Chavas, Jean-Paul., Michael Aliber, and Thomas L. Cox. 1997. An analysis of the source and nature of technical change: The case of US agriculture. *Review of Economics and Statistics* 79 (3): 482–492.

Ericksen, Polly J. 2008. Conceptualizing food systems for global environmental change research. *Global Environmental Change* 18 (1): 234–245. https://doi.org/10.1016/j.gloenvcha.2007.09.002.

Fanzo, Jessica, Lawrence Haddad, Rebecca McLaren, Quinn Marshall, Claire Davis, Anna Herforth, Andrew Jones, et al. 2020. The food systems dashboard is a new tool to inform better food policy. *Nature Food* 1 (5): 243–246. https://doi.org/10.1038/s43016-020-0077-y.

FAO. 2019. *The state of food and agriculture 2019: Moving forward on food loss and waste reduction.* Rome.

Global Panel on Agriculture and Food Systems for Nutrition (GloPan). 2016. *Food systems and diets: Facing the challenges of the 21st century.* London, UK.

Global Panel on Agriculture and Food Systems for Nutrition (GloPan). 2020. *Future Food Systems: For people, our planet, and prosperity.* London, UK.

Gundersen, Craig. 2019. The right to food in the United States: The role of the supplemental nutrition assistance program (SNAP). *American Journal of Agricultural Economics* 101 (5): 1328–1336. https://doi.org/10.1093/ajae/aaz040.

Haddad, Lawrence, Corinna Hawkes, Jeff Waage, Patrick Webb, Charles Godfray, and Camilla Toulmin. 2016. *Food systems and diets: Facing the challenges of the 21st century*. London, UK: Global Panel on Agriculture and Food Systems for Nutrition.

Herrero, Mario, Philip K. Thornton, Daniel Mason-D'Croz, Jeda Palmer, Tim G. Benton, Benjamin L. Bodirsky, Jessica R. Bogard, et al. 2020. Innovation can accelerate the transition towards a sustainable food system. *Nature Food* 1 (5): 266–272. https://doi.org/10.1038/s43016-020-0074-1.

Herrero, Mario, Philip K. Thornton, Daniel Mason-D'Croz, J. Palmer, B. L. Bodirsky, P. Pradhan, C. B. Barrett, T. G. Benton, et al. 2021. Articulating the impact of food systems innovation on the Sustainable Development Goals. *The Lancet Planetary Health* 5 (1): e50–e62. https://doi.org/10.1016/S2542-5196(20)30277-1.

High Level Panel of Experts on Food Security and Nutrition (HLPE). 2017. *Nutrition and food systems*. Rome.

High Level Panel of Experts on Food Security and Nutrition (HLPE). 2020. *Food security and nutrition: Building a global narrative towards 2030*. Rome.

Homer-Dixon, Thomas, Brian Walker, Reinette Biggs, Anne-Sophie Crépin, Carl Folke, Eric F Lambin, Garry D. Peterson, et al. 2015. Synchronous failure: The emerging causal architecture of global crisis. *Ecology and Society* 20. The Resilience Alliance. https://doi.org/10.5751/ES-07681-200306.

International Labour Organization (ILO). 2015. Agriculture: A hazardous work. Geneva: ILO. Accessed November 7, 2020. https://www.ilo.org/global/topics/safety-and-health-at-work/areasofwork/hazardous-work/WCMS_356550/lang--en/index.htm.

IPBES. 2019. Summary for policymakers of the global assessment report on biodiversity and ecosystem services of the Intergovernmental Science-Policy Platform on Biodiversity and Ecosystem Services. S. Díaz, J. Settele, E. S. Brondízio E.S., H. T. Ngo, M. Guèze, J. Agard, A. Arneth, P. Balvanera, K.A. Brauman, S.H.M. Butchart, K.M.A. Chan, L.A. Garibaldi, K. Ichii, J. Liu, S.M. Subramanian, G.F. Midgley, P. Miloslavich, Z. Molnár, D. Obura, A. Pfaff, S. Polasky, A. Purvis, J. Razzaque, B. Reyers, R. Roy Chowdhury, Y.J. Shin, I.J. Visseren-Hamakers, K.J. Willis, and C.N. Zayas, eds. IPBES secretariat, Bonn, Germany, 56 pages. https://doi.org/10.5281/zenodo.3553579.

IPCC. 2019. Climate change and land: An IPCC special report on climate change, desertification, land degradation, sustainable land management, food security, and greenhouse gas fluxes in terrestrial ecosystems. P.R. Shukla, J. Skea, E. Calvo Buendia, V. Masson-Delmotte, H.-O. Pörtner, D.C. Roberts, P. Zhai, R. Slade, S. Connors, R. van Diemen, M. Ferrat, E. Haughey, S. Luz, S. Neogi, M. Pathak, J. Petzold, J. Portugal Pereira, P. Vyas, E. Huntley, K. Kissick, M. Belkacemi, J. Malley, eds. https://www.ipcc.ch/srccl/.

Marshall, Quinn, Jessica Fanzo, Christopher B. Barrett, Andrew D. Jones, Anna Herforth, and Rebecca McLaren. 2021. Building a global food systems typology: A new tool for reducing complexity in food systems analysis. *Frontiers in Sustainable Food Systems* 5: 432. https://doi.org/10.3389/fsufs.2021.746512.

Maystadt, Jean-François., and Olivier Ecker. 2014. Extreme weather and civil war: Does drought fuel conflict in Somalia through livestock price shocks? *American Journal of Agricultural Economics* 96 (4): 1157–1182.

National Academies of Sciences, Engineering, and Medicine. 2020. *Innovation in the food system: Exploring the future of food: Proceedings of a workshop*. Washington, DC: National Academies Press. https://doi.org/10.17226/25523.

North, Douglass C. 1991. Institutions. *Journal of Economic Perspectives* 5 (1): 97–112.

Ridley, Matt. 2020. *How innovation works*. New York: Harper.

Timmer, C. Peter, Walter D. Falcon, and Scott R. Pearson. 1983. *Food Policy Analysis*. Baltimore, MD: The Johns Hopkins University Press for The World Bank.

The Economics of Ecosystems and Biodiversity (TEEB). 2018. *TEEB for Agriculture & Food: Scientific and Economic Foundations*. Geneva: UN Environment.

Vidar, Margret, Yoon Jee Kim, and Luisa Cruz. 2014. *Legal developments in the progressive realization of the right to adequate food*. FAO. http://www.fao.org/right-to-food/resources/resources-detail/en/c/271813/.

Von Uexkull, Nina, Mihai Croicu, Hanne Fjelde, and Halvard Buhaug. 2016. Civil conflict sensitivity to growing-season drought. *Proceedings of the National Academy of Sciences* 113 (44: 12391–12396.

Webb, Patrick, Tim G. Benton, John Beddington, Derek Flynn, Niamh M. Kelly, and Sandy M. Thomas. 2020. The urgency of food system transformation is now irrefutable. *Nature Food* 1 (10): 584–585. https://doi.org/10.1038/s43016-020-00161-0.

Willett, Walter, Johan Rockström, Brent Loken, Marco Springmann, Tim Lang, Sonja Vermeulen, Tara Garnett, et al. 2019. Food in the Anthropocene: The EAT–Lancet Commission on healthy diets from sustainable food systems. *The Lancet* 393 (10170): 447–492. https://doi.org/10.1016/S0140-6736(18)31788-4.

The State of Agri-Food Systems and Agri-Food Value Chains in 2020

One might reasonably invoke Dickens in describing AFSs and AVCs today: "it was the best of times, it was the worst of times." **There has been indisputable progress over the past hundred years, even the past decade. But there has also been backsliding, and contemporary AFSs are utterly unsustainable**, with massive, adverse spillover effects on the natural environment, public health, and social justice. Optimists and pessimists can each find support for their views in the data on contemporary AFSs.[1]

Remarkable agricultural productivity gains occurred over the past century, as exemplified by gains in maize (corn) yields in the United States (Fig. 1). But the agri-food research and development (R&D) that yielded these gains has been heavily concentrated in a small number of crops, primarily starchy cereals (e.g., maize, rice, and wheat), roots and tubers (e.g., potatoes), and livestock. This has led to declining relative prices of these staple commodities as compared to nutrient-rich fruits, legumes, nuts, and vegetables that have received far less R&D investment and which few countries produce in quantities sufficient to meet their populations' dietary requirements (Pingali 2012; Mason-D'Croz et al. 2019; Haddad 2020; Sanchez 2020).

Moreover, these productivity gains have also varied sharply across regions (Fuglie et al. 2020) and food system types (Fig. 2). We see variation in the magnitude of change, shown as longer time sequences in Fig. 2. Productivity gains in the world's industrial and consolidated AFSs have outpaced those of the rural and

[1] This is apparent in the recently released Food Systems Dashboard, which provides the most up-to-date data available on over 150 different indicators describing food systems at country, regional, and global scales (Fanzo et al. 2020).

C. B. Barrett et al., *Socio-Technical Innovation Bundles for Agri-Food Systems Transformation*, Sustainable Development Goals Series,
https://doi.org/10.1007/978-3-030-88802-2_2

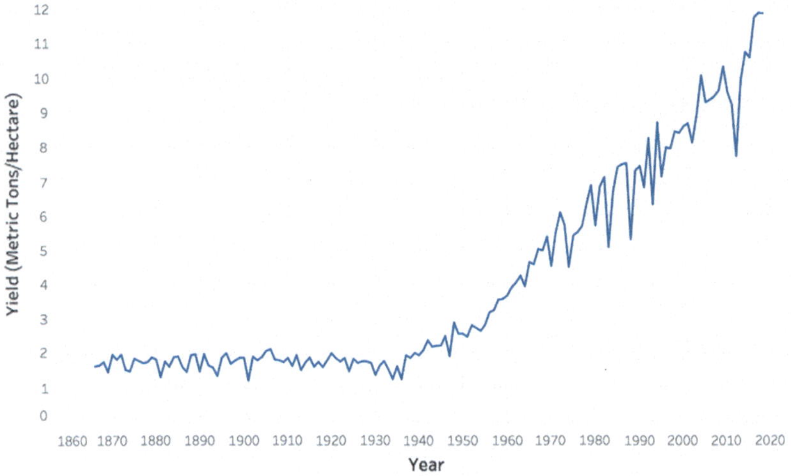

Fig. 1 Average maize (corn) yields in the United States, 1866–2014, in metric tons/hectare (*Source* United States Department of Agriculture and UN FAOSTAT)

traditional systems. Moreover, differences exist not only in the magnitude of productivity gains over time but also in their biases in favor of laborers or land owners. In rural and traditional systems (mostly the poorest regions of sub-Saharan Africa and South Asia), advances in improved germplasm, irrigation, etc., have mainly favored gains in land productivity (i.e., yield growth) that mainly benefit landowners. This is reflected in expansion curves that climb more steeply than the dashed, diagonal lines representing constant land/labor ratios in primary agricultural production. Conversely, labor productivity growth (e.g., from labor-saving machinery and agrochemicals) that chiefly rewards workers has outpaced land productivity growth in industrial and consolidated AFSs. Poverty remains both more pervasive and deeper in rural areas than urban ones in most of the world, coincident with the places where people depend most heavily on AVCs for their livelihoods as farmers, farm workers, transporters, meatpackers, etc.

Figure 2 also plainly reveals the stark difference in productivity across AFSs. Agricultural output per unit land in production is severalfold higher in industrialized systems than in traditional ones—reflecting the crop yield gaps on which so much of the agricultural sciences community focuses. But these gaps pale in comparison to those in labor productivity. Agricultural output per adult employed in agriculture is nearly two orders of magnitude greater in the industrialized

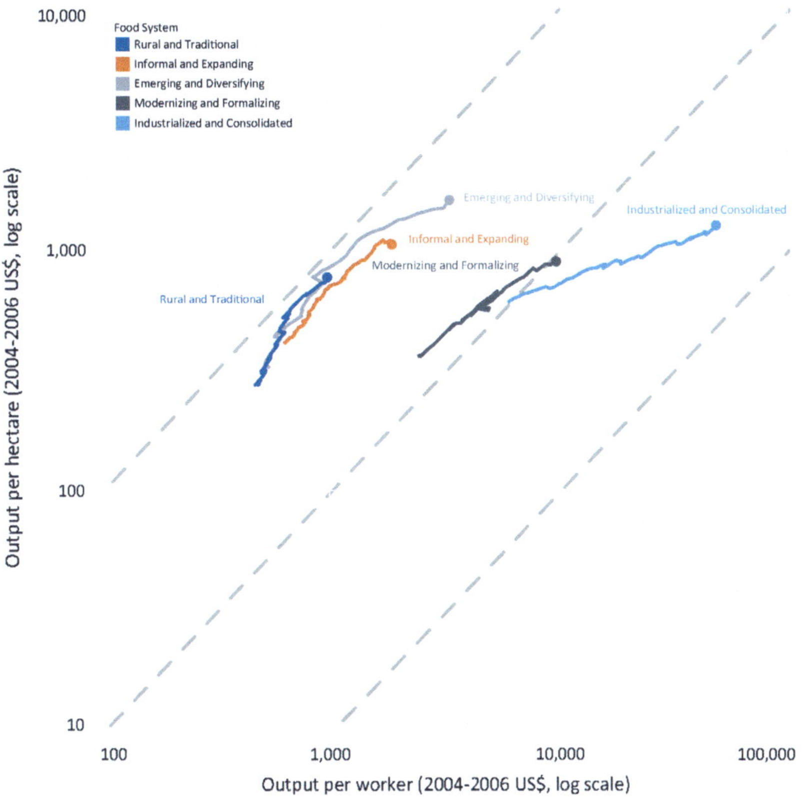

Fig. 2 Trends in agricultural land and labor productivity, 1961–2016, by food system type. Colored lines show changes in productivity over time, from 1961 through 2016. Output is in 2004–2006 international dollars. Labor reflects number of adults employed in agriculture, and land as agricultural land in rainfed equivalent (*Data source* USDA-ERS International Agricultural Productivity Database; figure adapted from Fuglie et al. 2020)

systems than in the traditional ones. This stark difference is a central reason for radical differences in living standards across the globe. Many technologies and practices already widely in use could significantly close those gaps,[2] but for

[2]As but one example, on-farm experiments in Nigeria generated dramatic yield gains in cassava simply through generous fertilizer application (Adiele et al. 2020).

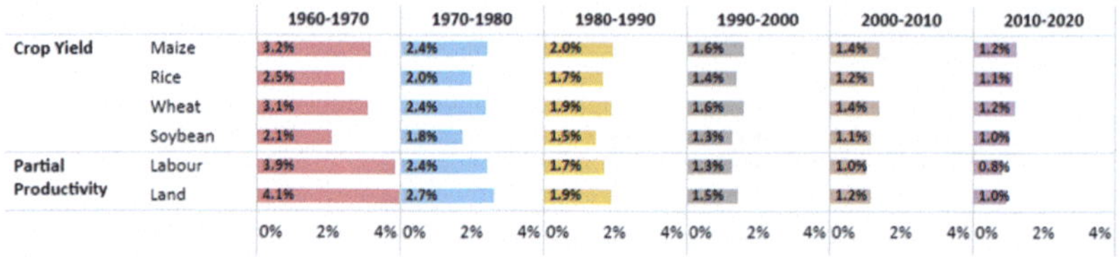

Fig. 3 Global crop yield, labor, and land productivity annualized growth rates, 1960–2020. Estimated as compound annual growth rates per decade, based on regressions of global data (*Data sources* FAOSTAT for crop yields and USDA-ERA International Agricultural Productivity Database for partial productivity measures)

myriad reasons are not widely available or adopted in poor rural areas in the low-income world. Closing existing productivity gaps using extant knowledge could help advance equity and healthy diets goals quickly, but too often with significant environmental and climate sustainability tradeoffs.

The rate of agricultural productivity growth has slowed markedly over the last generation, however (Alston et al. 2009; Fuglie et al. 2020, Fig. 3). In addition, agri-food R&D has increasingly shifted to the private sector. Private R&D now accounts for more than two-thirds of total agricultural R&D spending in both China and the US (Chai et al. 2019). One result is that intellectual property rights (e.g., patents) are increasingly likely to impede affordable access to, and adaptation of, new discoveries. Partly as a result, the R&D cost per unit productivity gain has also been rising rapidly (Bloom et al. 2020). The gap between high- and low-income country agri-food R&D has been growing (Pardey et al. 2016). Meanwhile, anthropogenic climate change has countered some of the favorable impacts of technological change, reducing global agricultural total factor productivity growth by 21 percent since 1961, equivalent to losing roughly a decade's productivity growth (Ortiz-Bobea et al. 2020).

Gains in on-farm productivity have helped propel growth downstream in food processing and distribution. AVCs continue to dominate employment, especially in poorer countries. The agricultural share of an economy's labor force steadily declines as part of the inevitable process of structural transformation, in which workers migrate from agriculture to other sectors even as agricultural output grows and despite agriculture's greater labor-intensity than other economic sectors (Barrett et al. 2017; Mellor 2017). But growth in downstream portions of AVCs accelerates at the same time. Today, employment in the post-harvest segments of AVCs dwarfs on-farm jobs and is growing globally, even by a factor of ten in Sub-Saharan Africa (Thurlow 2020; Yi et al. 2021; Dolislager et al. 2021).

While AVCs employ more people worldwide than any other sector—more than 1.3 billion (ILO 2015)—**AVC jobs are also more poorly compensated, dangerous, and precarious than those in any other sector save mining, and more prone to child, forced, and unsafe labor than those in any other sector but textiles**. The International Labour Organization (ILO) reports that agricultural workers account for approximately half of all fatal occupational accidents annually (ILO 2017). Marginalization and group-based discrimination—against women, ethnic, racial, or religious groups, etc.—is pervasive in AVCs. This marginalization typically reflects broader systemic discrimination within the societies of which AVCs are a part. These features intersect, as economic desperation and sociopolitical marginalization drive under-resourced groups to take on more perilous and poorly compensated work. The concentration of marginalized populations in AVC employment that is more dangerous and less remunerative than employment in other sectors thus magnifies broader societal problems within AFSs. Partly as a result, smallholder, farm, and AVC worker households are disproportionately likely to suffer food insecurity (FAO 2019).

Over the past 30 years, science-based advances in AFSs have boosted both food supplies and incomes. This has enabled an average of 90 million additional people each year to secure at least minimally adequate daily dietary energy intake (Fig. 4). But since 2014, and even prior to the 2020 pandemic, the number of undernourished and the prevalence of moderate and severe food insecurity have been slowly increasing, even as the total population that is food secure and receives adequate dietary energy intake has also increased (due to population growth). The undernourished increasingly concentrate in conflict-affected countries.

Past AFS advances were not designed with fragile settings in mind, thus different tools are increasingly needed to address hunger and famine concerns that are closely bound up with conflict (Barrett 2021). Today at least 3 billion people cannot afford a healthy diet, the cost of which exceeds the international poverty line, with dietary shortfalls especially concentrated among essential minerals and vitamins (FAO 2019). On the flip side, never before have more than 4.5 billion people been able to afford and consume a healthy diet (Barrett 2021)—once again, both the best of times and the worst of times.

Sustained productivity growth in AFSs drove real (i.e., inflation-adjusted) food prices to all-time lows at the turn of the millennium. And consumer food-budget shares have continued to decrease thanks to real income growth, especially in emerging markets in Africa and Asia. But real food prices both rose significantly and became more volatile over the first two decades of the twenty-first century (Fig. 5).

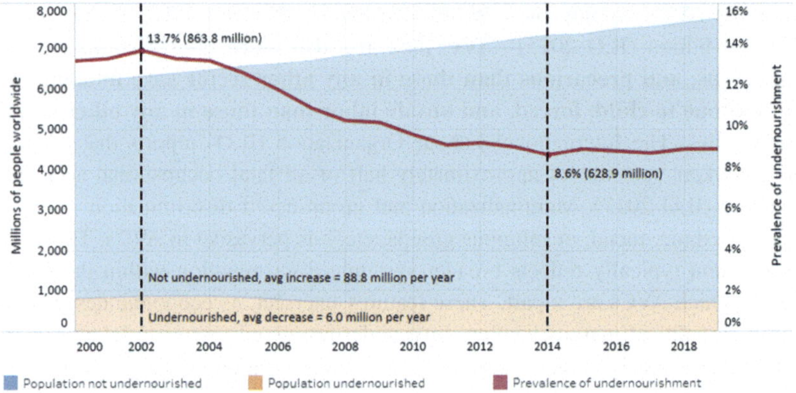

Fig. 4 Global population undernourished, 2000–2019. The colored areas reflect the number of people (not) undernourished (blue and yellow, respectively). The red line shows the global prevalence of undernourishment. The vertical dashed lines reflect the high and low points this century for prevalence, with the associated number of undernourished in parentheses (*Data source* FAOSTAT)

Ironically, the human population and income growth that now challenge sustainable management of natural systems and help foster a global overweight and obesity public health crisis have been enabled by scientific discovery that made food cheaper (Fogel 2004; Barrett 2021). **Cheaper calories and protein have naturally led to massive dietary change, and not all for the better** (Fig. 6). **Diet is now the top risk factor for morbidity and mortality globally** (GBD 2019), as per capita daily consumption of meats, empty calories (refined sugars, refined animal fats, oils, alcohol), and total calories have increased dramatically over time but also quite unevenly across country groups. As processed products[3] represent an ever-growing share of what food consumers eat, the challenges of

[3] There is no universally accepted definition of processed foods. The basic idea is that processed foods have undergone one or more changes to their natural, raw commodity state. That may involve blanching, canning, cooking, dehydrating, drying, freezing, milling, washing, etc., as well as combination in manufacturing that uses processed foods as inputs. Ultra-processed or "highly processed" foods are another ambiguous term, by which one typically means foods that have added fats, salt, or sweeteners and/or artificial colors, flavors, or preservatives, with the objective of promoting shelf stability or palatability, or preserving texture, but often at a cost of decreased healthfulness in some dimension.

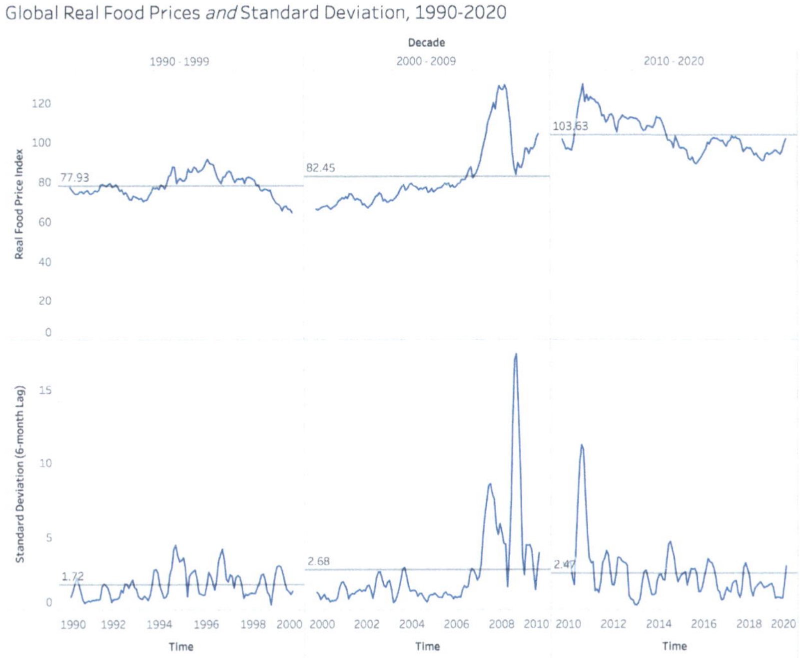

Fig. 5 Global real food prices, January 1990–July 2020 (*Data source* FAO Food Price Index)

inducing higher-quality processing and more healthy (re)formulation loom larger than ever. Not all processed foods are unhealthy, although the market and regulatory incentives presently facing food manufacturers and food service firms such as restaurants broadly favor low-cost, unhealthy refined sugars and fats.

Further, the considerable food loss and waste in today's AFSs—FAO (2019) estimates 14 percent average loss post-harvest, not including retail/consumer waste—are partly a direct function of cheap food (FAO 2019; Cattaneo et al. 2020). United Kingdom households, for example, waste the equivalent of 42 daily diets per capita per year, on average, with significant losses of key nutrients already deficient in the diet (Cooper et al. 2018).[4] Indeed, for some essential

[4]The total estimated climate impact was 20.4 million tons CO_2-equivalent per year, roughly comparable to 6.5 million round trips across the United States by car.

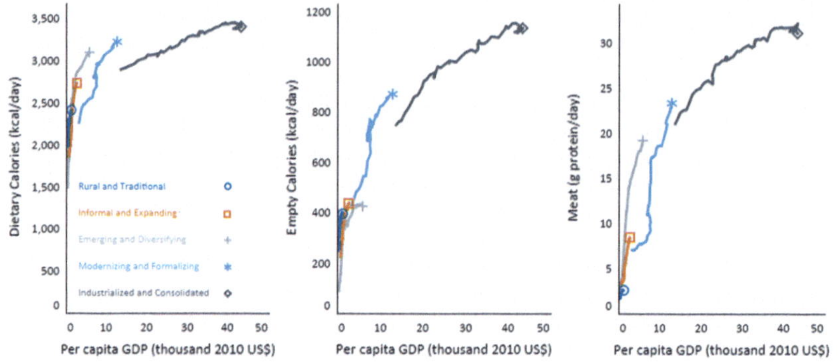

Fig. 6 Shifting food consumption patterns with income growth. Colored lines reflect 1961–2013 average consumption trends with respect to per capita gross domestic product (GDP) in thousand 2010 US dollars. Empty calories estimated as calories from sugars, sweeteners, vegetable oils, and alcohol (*Data sources* FAOSTAT for calories and protein, World Bank World Development Indicators for per capita GDP; figure adapted from Tilman and Clark 2014)

nutrients, such as calcium or folate, residual nutrient availability after accounting for global loss and waste is less than 10 percent above the recommended daily dietary requirements (Ritchie et al. 2018), implying massive prevalence of micronutrient deficiencies given the grossly inequitable distribution of healthy foods across the global population. While food loss and waste is generally considered from a "farm to fork" perspective, the disposal of post-consumption nutrients (through sanitary services or otherwise) can also be regarded as a form of waste, with enormous environmental and health consequences. "Fork to farm" approaches that recover resources for agriculture can address sanitation, health, and food security challenges, as discussed below.

Innovations in plant and animal genetics and nutrition, irrigation, mechanization, and other technologies have enabled the intensification of production to an extent that has obviated massive amounts of deforestation (Evenson and Gollin 2003; Pelletier et al. 2020; Gollin et al. 2018). But "modern agriculture" has depended heavily on dramatically increased use of inputs, including nitrogenous fertilizers made with the heavy use of petrochemicals, mined phosphates, irrigation, and pesticides (Tilman et al. 2002). Each of these input types are associated with problems and concerns related to environmental sustainability, as we discuss below.

Rural lands have massive potential to sequester carbon in soils and trees but today are a major source of avoidable GHG (i.e., CO_2, CH_4, and N_2O)

emissions.[5] Incentives based on production, global competition based on price, and long supply chains reducing transparency encourage the externalization of significant costs on the environment. This includes impacts on:

- Soils and their degradation through compaction, loss of organic carbon, salinization, and erosion (Amundson et al. 2015).
- Biodiversity, where AVCs are the biggest driver of biodiversity loss (Newbold et al. 2016; IPBES 2019).
- Water, where extraction may reduce water below the safe level for environmental integrity and deplete aquifers, as well as impact water quality through various forms of agricultural run-off. Nutrients in run-off have adverse consequences, contributing to harmful algal blooms, dead zones affecting coastal fisheries, disease outbreaks, and other environmental and human health issues (Dalin et al. 2017; Kanter et al. 2020).
- Air quality, which is affected by the use of fertilizers and the burning of fossil fuels and crops residues. (As an example of the scale of the issue, one locational study suggested that the health-related costs of agriculture are approximately half the value of the agriculture itself [Paulot and Jacob 2014].)
- The concentration of GHGs, which are a major driver of climate change. (AVCs emit as much as 30 percent of anthropogenic GHGs [Bajželj et al. 2013; Poore and Nemecek 2018].)

The per capita environmental footprint of AVCs is significant. Each global citizen's AVC use averaged about three-quarters of a hectare of land (Davis et al. 2016); 776 tons of water, typically mostly rainwater (Davis et al. 2016); 284 g of pesticide-active ingredient (FAOSTAT as of 2015); 9 g of antimicrobials (van Boeckel et al. 2015); and 15 kg of nitrogen fertilizer (Davis et al. 2016), while at the same time emitting just over 2000 kg of CO_2 equivalent (IPCC 2019).

Concerns about deteriorating resilience to growing risks abound. The number of natural disasters worldwide has been increasingly steadily, up more than threefold from 1980 to 2019, with most associated losses uninsured, especially in the low- and middle-income countries (LMICs) where insurance coverage is less than 10 percent (Munich Reinsurance 2020). Massive shocks that disrupt agricultural production more specifically (e.g., droughts, flooding, deadly tropical storms,

[5]We note, however, that soil carbon sequestration capacity diminishes as soils saturate, while tree growth's sequestration potential does not taper as much, if at all. Both are, however, reversible with changes in soil and forest management practices.

locusts, fall armyworm, and other pests) have, likewise, grown in frequency, severity, and potential for co-occurrence with other shocks that compound damages. The COVID-19 pandemic is unlikely to be the last one of this century, so learning lessons from the massive disruptions of 2020 will be imperative to building back better and more resilient in the future. Largely due to war, but increasingly due to climate change, according to the United Nations High Commissioner for Refugees, 80 million forcibly displaced people had fled their homelands at the end of 2019, more than at any time since World War II (UNHCR 2020). Addressing humanitarian needs is far more costly in both human and financial terms the further people move from their homes.

Nonetheless, the scope for AFS changes to reduce hunger and acute malnutrition grow increasingly limited. The reason is that outside of zones of active, violent conflict (e.g., Yemen currently; Somalia, especially in 2011; or South Sudan, Northeast Nigeria, and eastern Democratic Republic of the Congo episodically over the past decade) and states with severe governance problems (e.g., North Korea or Venezuela) famine and near-famine conditions have largely disappeared with advances in early warning systems and humanitarian response, greater inter-regional market integration, and more inclusive and effective social protection programs (Alderman et al. 2017; Maxwell and Hailey 2020). The acute malnutrition and chronic hunger problems that motivated the last concerted global efforts at AFS transformation in the 1960s and 1970s have become primarily problems of conflict resolution and humanitarian response (Barrett 2021).

The growing link between acute malnutrition and humanitarian response, together with heightened concerns of fragility in key tropical ecosystems, have rapidly drawn attention to broad questions of resilience (Barrett and Constas 2014; Hoddinott 2014; Tendall et al. 2015; Béné 2020). Resilience encompasses notions of resistance to, and recovery from, shocks. Will a shock perturb food supply or access to food? If so, how great a perturbation will occur, and how quickly and closely will it return to—or improve upon—previous functionality?

Resilience, whether at the production level or at the food system level, typically arises through one or both of two mechanisms: functional redundancy and diversity. The first typically would arise from having spare capacity (e.g., food stores for supplies, or decentralized processing so that there is no single point of failure). The second would include diversity in food products, suppliers, geographies, and products (e.g., multiple crop varieties/species or animal breeds/species so that a stress is less likely to hit at the most vulnerable point for all species). Both notions typically run antithetical to standard "efficiency" considerations, which rely on monocultures optimized for typical conditions and just-in-time supply chains that engage preferred suppliers who are highly specialized with no scope for substitution of products.

Building resilience, therefore, almost inevitably requires incurring additional costs relative to the way well-resourced AFSs have evolved under intense uninsured cost-minimization pressures from short-run profit-minded companies and investors. Socially optimal pricing must build in the cost of insurance against catastrophic shocks. Companies that embrace the transformational changes required and undertake appropriately ambitious actions recognize this risk and can ensure the appropriate long-term thinking and funding to enable the needed changes. When made part of a company's purpose, this reorientation has proved capable of attracting like-minded investors, as well as having beneficial impacts on other factors such as employee retention and brand loyalty.

BUILDING RESILIENCE, THEREFORE, ALMOST INEVITABLY REQUIRES INCURRING ADDITIONAL COSTS RELATIVE TO THE WAY WELL-RESOURCED AGRI-FOOD SYSTEMS HAVE EVOLVED UNDER INTENSE UNINSURED COST-MINIMIZATION PRESSURES FROM SHORT-RUN PROFIT-MINDED COMPANIES AND INVESTORS.

From an ecological perspective, AVCs typically reduce ecological resilience by reducing diversity. Agriculture modifies landscapes from small to large scales in multiple ways, typically creating homogeneity at scale (Benton et al. 2003). As a result, across about two-thirds of the Earth's land surface, ecological communities have been radically affected (Newbold et al. 2016). "Modern" agriculture commonly creates input-intensive monocultures by amalgamating small parcels of land into large, uniform blocks, accelerating the decline of both agricultural and wild biodiversity (Kremen and Miles 2012). Actively removing heterogeneity in the environment leaves the world vulnerable to pathogens and pests that can decimate crops at massive scale (Fones et al. 2020), depletes beneficial soil microbial communities (Zhao et al. 2018; Tang et al. 2019), and can allow weed communities to thrive (Poggio 2005) partially due to soil nutrient depletion occurring under uniform cropping patterns (Ehrmann and Ritz 2014).

Modern agriculture increasingly relies on inputs that have direct effects that bost farm productivity (e.g., pesticides killing pests) but which also kill "non-target organisms" (e.g., non-pests which may be the natural enemies of pests) and adversely spill over to other habitats, while also depending on fertilizers that negatively affect air and water quality. Large-scale enterprises can achieve efficiencies of scale and scope that boost conventional economic measures of total factor productivity but concentrate adverse impacts, as when intensive, large-scale livestock enterprises create mass manure lagoons that are difficult to manage and risk catastrophic damage to nearby watersheds. Habitat complexity on a local scale is particularly important for maintaining specialist predator populations that are important for pest control (Chaplin-Kramer et al. 2011).

Lessons from the COVID-19 Pandemic: Directing Inevitable AVC Innovation[6]

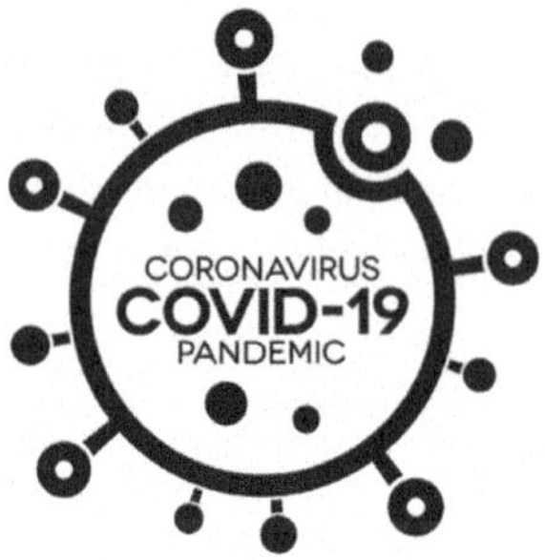

The COVID-19 pandemic serves as both a warning and an accelerator. As above, the data support both optimistic and pessimistic interpretations, revealing strengths, vulnerabilities, and weaknesses of modern AFSs. The pandemic has also underscored that simply returning to what was previously "normal" will not be good enough.

Massive disruptions within AVCs have been commonplace throughout history. But most prior disruptions have been driven by supply-side shocks arising from a crop failure, a livestock disease outbreak, etc. In such cases, downstream buyers responded by finding other suppliers or drawing down stored commodities, bidding up prices temporarily until supply recovered. But in the COVID-19 pandemic, supply-side shocks have been largely restricted to some (relatively modest) labor supply disruptions, especially in Europe and India, arising from some nations' restrictions on worker migration or due to disease outbreaks in sites where workers operate in very close proximity to one another (e.g., slaughterhouses in Brazil and the US or at fruit and vegetable packing factories). Overall, primary production has proved remarkably robust. Indeed, the Food and Agriculture Organization (FAO) forecasts record global harvests for 2020.

The world has likewise grown accustomed to isolated logistics disruptions associated with natural disasters (e.g., floods or earthquakes that knock out roads or bridges) or war and other forms of violence that disrupt the flow of food and drive up costs in specific, disaster-affected regions. Despite food export bans—most of them lasting only a few weeks—imposed by at least 20 different national governments (Laborde et al. 2020) and massive shutdown of commercial passenger transport, merchandise freight shipments have been largely untouched, especially in multinational firms' global supply chains. Virtually all AVCs recovered reasonably quickly to supply-side and logistics-driven disruptions associated with COVID-19.

[6] Icon courtesy of Covid Vectors by Vecteezy (https://www.vecteezy.com/free-vector/covid). A revised version of this sub-section appeared as Barrett et al. (2021).

The damage to AVCs from the COVID-19 pandemic, for the first time in living memory, occurred overwhelmingly from a massive demand-side shock to AVCs, as widespread closure of many businesses (disproportionately food service operations—both commercial ones like restaurants or entertainment venues, and institutional ones such as school cafeterias) left hundreds of millions of people worldwide suddenly without jobs and the income to acquire a healthy diet (Barrett 2020). The loss of livelihoods has nearly doubled the number of people worldwide suffering acute food insecurity, to an estimated 270 million.[7] This sparked long lines for private food assistance and sharp expansion of public food assistance.

Meanwhile, **food service accounts for a large and growing share of food consumption globally—roughly half of all consumer food expenditures in high-income countries**—so the pandemic represented a massive disruption to AVCs structured around serving people food away from home. The unprecedentedly fast and severe economic shock induced panic buying as food consumers were forced to redirect virtually all of their demand towards retail outlets. The shuttering of food service enterprises and resulting shutdown of value chains built to deliver to those outlets caught many farmers and food manufacturers with unsellable perishable products. Livestock farmers were effectively compelled to euthanize animals and to dump milk and eggs into waste lagoons. Horticultural producers plowed ripe fruits and vegetables back into their fields. And manufacturers ran out of warehouse storage space for bulk processed goods packaged for institutional buyers.

The most common responses by governments and private charitable organizations have been (1) public health measures to control and treat COVID-19, and (2) unprecedented expansion of safety net and social protection programs (Gentilini et al. 2020). The mechanisms for doing so have varied markedly across, and within, countries—from universal basic income programs, to employment guarantee schemes, government payroll subsidy programs, enhanced unemployment insurance, and expanded access to food assistance. In the short interval of March–September 2020, 212 different countries/territories announced and/or implemented an astounding 1,179 different social protection measures in response to the massive dislocations caused by the COVID-19 pandemic (Gentilini et al. 2020). The necessity of supporting consumer demand, especially

[7] Per the UN World Food Programme estimates from June 2020 (https://www.wfp.org/news/world-food-programme-assist-largest-number-hungry-people-ever-coronavirus-devastates-poor).

among the poorest and most vulnerable, has been the centerpiece of societal response, not only to the pandemic in general but also to cushioning AFSs from the demand shock.

Overall, AVC intermediaries adapted quickly, switching among value chains and service modes. Restaurants quickly flipped to delivery, takeout, and outdoor dining options. Processors modified manufacturing processes to expand retail-oriented packaging while reducing wholesale packaging for food service clients.

Some of these adaptations are likely to prove permanent, as the pandemic boosted consumers' and companies' awareness of the value chains on which they draw, and farmers have become more aware of what happens downstream after they sell their product. This awareness has accelerated change towards online grocery purchases and food delivery, community-supported agriculture and similar direct-to-consumer arrangements, and home gardens. Ventures such as Malaysia's Myfishman.com, which connects fishermen to individual consumers, have flourished worldwide while communities have revived gleaning as a way to reduce food loss and improve poor consumers' access to healthy fresh foods.[8]

Already-growing demand for plant-based meat substitutes has likewise increased as consumers grew more concerned about the sustainability of production systems and the potential for food contamination in long value chains (Siegrist and Hartmann 2019; Van Loo et al. 2020; Jalil et al. 2020).[9] Crop and dairy farms, meatpackers, and other AVC firms have sharply stepped up investment in robots invulnerable to infectious disease transmission. Farmers, traders, manufacturers, and food service vendors have rapidly expanded their use of e-commerce platforms to help find customers and suppliers. Farmers and processors have adopted creative approaches to improve worker safety and firm resil-

[8] Gleaning is a centuries-old tradition of mobilizing small groups to collect edible crop left in the field after a harvest, or of unsellable crops left in the field. In the US, for example, 6–7 percent of planted acreage is unharvested because of cosmetic blemishes, mechanical harvesting error, or a lack of market for the crop (https://www.nytimes.com/2020/07/06/dining/gleaners-farm-food-waste.html; https://www.sciencedirect.com/science/article/pii/S0306919216301026).

[9] For example, Impossible Foods expanded its retail distribution of plant-based beef substitutes from less than 200 stores in January 2020 to more than 3000 stores by May 2020 (Nierenberg, *Wall Street Journal*, May 22, 2020), while Beyond Meat's revenue increased 69 percent year-on-year to June 2020 (Maidenberg, *Wall Street Journal*, August 4, 2020). See also Shahbandeh (2020, https://www.statista.com/topics/6057/meat-substitutes-market-in-the-us/). The global plant-based meat market is predicted to exceed US$35 billion by 2027 (Polaris 2020).

ience, such as the Nigerian chicken processors who organized dedicated bus transport for workers on more sparsely staffed shifts at factories (Reardon and Swinnen 2020). Meanwhile, governments and charitable organizations have doubled down on the use of mobile digital transfers of cash and vouchers for food assistance. Many of these changes are welcome advances unlikely to reverse once the health scare and economic dislocation of the pandemic passes.

The pandemic has also laid bare great structural inequities of risk exposure within AFSs. In high-income countries, "essential" workers in grocery stores, food delivery services, densely-packed meatpacking plants, etc., suffered far higher rates of infection and death than the food consumers they serve or white-collar executives in those same sectors. Essential workers were more likely to be people of color, not to have graduated from university, and to have lower income—all strong correlates of obesity and diet-related non-communicable diseases such as diabetes and hypertension. Those structural inequities existed long before the pandemic but have been magnified by it. More than a century after Upton Sinclair's *The Jungle* called attention to the inhumane working conditions in meatpacking plants, a groundswell of concern has reemerged about protecting farmworkers and meatpackers, both for their benefit and so as to safeguard food supplies and stem disease transmission from workers who migrate to follow harvest periods.

The COVID-19 pandemic has made clear that **healthfulness, equity, resilience, and sustainability are interlinked, precompetitive issues. They concern our collective fitness as a species when faced with covariate shocks** like pandemics, climate change, and mass extinctions. And this is a centerpiece of the challenge before us. Incentives that skew excessively towards the promotion of individual interests can undermine collective action (Ostrom 2010). Then virtually everyone is worse off because, as elementary game theory makes clear, cooperative outcomes are almost always superior to noncooperative ones, but cooperation typically arises only when the rules of the game naturally induce a critical mass of people to do so.

Trust underpins cooperation (Barrett 1997; Ostrom 2010). The pandemic has made clear the importance of cultural and political responses to scientific uncertainty and trust in expert guidance. Responses have varied wildly across, and within, countries. If cooperation is the watchword on precompetitive issues, then many communities have failed this recent, lethal test, as basic public health measures became deeply politicized. The pandemic is a trial run not just for inevitable, future infectious disease outbreaks, but also for climate change and biodiversity loss. These are, likewise, natural processes but with even larger-scale and longer-lasting implications for humanity and the AFSs that support us than that of COVID-19. As societies impose major sacrifices on younger generations in order to protect more vulnerable older populations, will reciprocity emerge wherein the older adults, who exercise most power in economic and political systems, accept responsibility to make some near-term sacrifices as investments to protect today's young and as-yet-unborn generations from avoidable ravages of climate change?

THE COVID-19 PANDEMIC HAS MADE CLEAR THAT HEALTHFULNESS, EQUITY, RESILIENCE, AND SUSTAINABILITY ARE INTERLINKED, PRECOMPETITIVE ISSUES. THEY CONCERN OUR COLLECTIVE FITNESS AS A SPECIES WHEN FACED WITH COVARIATE SHOCKS LIKE PANDEMICS, CLIMATE CHANGE, AND MASS EXTINCTIONS.

Even as science has become further politicized in some places during the pandemic, we have witnessed historically unprecedented mobilization of finance for basic and applied science to seek vaccines to prevent, and treatments for, COVID-19. Creative arrangements have emerged—not just conventional research contracts and grants to research institutions, or venture capital, conventional debt or equity financing of private laboratories, but also advanced market commitments to ensure a large-scale, remunerative commercial market necessary to induce private investment while simultaneously ensuring widespread access in low-income countries (GAVI 2020; Kremer et al. 2020).

The intellectual property behind whatever successful discoveries emerge will inevitably be hotly contested within, and among, countries. Pre-existing patents have not, however, impeded R&D progress, which has advanced at an unprecedented pace. Before COVID, the fastest vaccine ever developed, against mumps, took four years from initial sample collection and identification until vaccines were licensed for approved distribution. As this report goes to press, we appear on the cusp of vaccine approvals in just months, well under a year since the virus was first identified! The astounding pace of progress seems partly due to the Open-COVID Pledge launched in April 2020, which enables biomedical researchers to freely share their IP following a model similar to that used for open-source software; the pledge covered more than 250,000 patents worldwide by end-July (Contreras et al. 2020). **The COVID-19 experience clearly demonstrates that massive amounts of financing, scientific talent, and popular support can be mobilized quickly with adequate political will and a shared sense of urgency**, which are equally needed for the task of AFS transformation.

Mainly, the pandemic has been a wake-up call to prepare and build back better. The unprecedented global scale and speed of this shock to AFSs compel change. Return to the *status quo ex ante* seems both unlikely and unwise. At a defining moment when paths will almost-inevitably shift, we must focus intently on crafting innovation pathways that can effectively navigate the world from its current vulnerable condition to our desired states. The pandemic creates an opportunity to address systemic needs arising from other pressures (e.g., climate change) but to which the world has, to date, been insufficiently responsive. This can be a moment of "creative destruction," to invoke Joseph Schumpeter's famous term (Schumpeter 1942), a moment for dismantling established processes that cannot possibly deliver healthy diets, equitable and inclusive livelihoods, environmental sustainability, and resilience to shocks and stressors, and to replace them in a dynamic process of innovation and adaptive management. The following **thirteen key**, general lessons for AFSs stand out from the COVID-19 pandemic experience:

Stuff happens... be ready. This isn't a one-off, short-run shock. No sensible person believes this pandemic will be the last major challenge of our lifetimes. We must be prepared for more severe and more frequent, compound shocks, as well as for simultaneous and cascading shocks. This implies we need greater redundancy and resilience in AFSs and AVCs.

Expect that ever-ready social safety nets are needed. The pandemic's pain has aggravated underlying inequalities. Nations and communities need reliable, scalable social protection programs that are sensitive to race, gender, ethnicity, and other dimensions of systemic discrimination. These cannot be built on the fly. Weak or incomplete social protection mechanisms undermine solidarity and cooperation within society, thereby discouraging responsible individual behaviors and hurting everyone.

Beware slower-moving catastrophes. The pandemic was fast-moving, compelling policymaker attention. We must beware slower-moving—but no less consequential—shocks, such as those due to climate change, biodiversity and habitat loss, sea level rise, etc. Slower transition can engender complacency—the mythical frog-in-the-water-as-it-warms problem—and can imply lesser ability to get the shock under control once people finally feel compelled to act.

Realize that massive resources can be mobilized quickly. Trillions of dollars have been appropriated by governments in just a few months. Where the needs are apparent and political leaders feel compelled to act, funds can be found fast (Herrero and Thornton 2020).

Move beyond uninsured cost minimization. Affordable, healthy diets are crucial for equity purposes but often involve resilience and sustainability tradeoffs. **De-risking AFSs requires greater diversification of production, sourcing, processing, and distribution patterns to enhance flexibility and redundancy**. This has a cost but also a value, as costly insurance against catastrophic systemic risk always does.

Beware de-globalization. Supply chain disruptions have fueled many governments to pursue food self-sufficiency more aggressively. This carries significant prospective risk. First, **de-globalization can harm the poor by making healthy diets more expensive**. Second, it can undermine environmental and climate sustainability because *how* a product is produced, processed, and distributed matters far more to its footprint than *where* it was made (Poore and Nemecek 2018). Third, trade is essential to manage changing climate (Baldos and Hertel 2015). Fourth, the more countries disengage from one another and pursue trade wars, the greater the likelihood of interstate conflict, which is the single greatest cause of severe acute malnutrition globally (Barrett 2013). Build more diversified and resilient AVCs, but be careful about hidden nationalist agendas.

Fund and trust first-rate science. Technical skill is essential preparation. We can adaptively manage and innovate only if we can learn fast. **We cannot build scientific and engineering capacity overnight but can undermine it quickly through poor communications, especially if leaders let politics overrule, and even misrepresent science**.

Understand that barriers to success are more behavioral than scientific. Although the science on COVID-19 has progressed at unprecedented speed, behavioral adjustments have proved far slower and more uneven across communities. Culture change is key and requires convincing social influencers and thought leaders to do things differently as we learn. This also requires checks and balances to avoid excessive concentration of political/commercial power, which has strong conservative tendencies to entrench itself.

Recognize that clear, consistent, trusted incentives and norms are key. No coordinated response emerged at global scale and not even at national scale in most countries. **The enormous numbers of independent agents throughout centralized AVCs made market incentives and social norms, not top-down directives other than to drive incentives and calibrate norms, the key policy instruments**. Decentralized, market-based AVCs self-stabilized reasonably quickly and well under the circumstances, especially where markets were allowed to induce rapid response to shutdowns in AVC subsectors.

Value communication, transparency, and cooperation as essential. Spillovers are ever-present, so strong coordinating institutions are essential to build and maintain trust so as to quickly identify and contain contagion. Because trust inevitably requires verification, traceability is increasingly at a premium.

Assume that dramatic, fast improvements are possible. Behavioral change is hard but feasible. Societies worldwide rapidly adjusted, virtually shutting sectors (e.g., food service, commercial transport). This generated sharp reduction in disease transmission and in GHG and pollution emissions. These results demonstrate clearly that we can dramatically improve outcomes if we have the incentives to exert ourselves.

Treat underlying causes, not just symptoms. Pandemics are the long-predicted consequence of habitat and biodiversity loss (partly due to expanding land use in agriculture) that increases exposure to zoonoses, of inconsistent and non-transparent food safety regulations, and of insufficient integration between food and health systems. Root cause analysis is key to ensure each limiting factor is identified.

Emphasize high-frequency monitoring. Systemic shocks require near-real-time monitoring of fast-changing conditions. Innovations in remote sensing, digital records, "sewage epidemiology" (monitoring biomarkers for disease and other exposures in human and animal waste streams), and crowd-sourcing open up new opportunities to improve the timeliness and cost-effectiveness of responses to systemic shocks.

Crises inevitably spark innovation. The crucial questions are what sorts of innovation will happen as AVCs recover from the COVID-19 pandemic, and how can we best induce beneficial innovations? Because a disproportionate share of the reconstruction of AVCs will—and must—happen in the coming 2–5 years, near-term innovations—in institutions and policies, as much as in technologies—will likely lock in for some time as investors and policymakers amortize the sunk costs they incur. So we need to influence today's innovations with an eye to decades hence. What should the design objectives be, and what will AFSs and AVCs look like in 25–50 years (i.e., the lifespan of a current person of median global age)?

References

Adiele, J.G., A.G.T. Schut, R.P.M. van den Beuken, K.S. Ezui, P. Pypers, A.O. Ano, C.N. Egesi, and K.E. Giller. 2020. Towards closing cassava yield gap in West Africa: Agronomic efficiency and storage root yield responses to NPK fertilizers. *Field Crops Research* 253: 107820.

Alderman, Harold, Ugo Gentilini, and Ruslan Yemtsov. 2017. *The 1.5 billion people question: Food, vouchers, or cash transfers?* Washington, DC: World Bank. https://doi.org/10.1596/978-1-4648-1087-9.

Alston, Julian M., Jason M. Beddow, and Philip G. Pardey. 2009. Agricultural research, productivity, and food prices in the long run. *Science* 325 (5945): 1209–1210. https://doi.org/10.1126/science.1170451.

Amundson, Ronald, Asmeret Asefaw Berhe, Jan W. Hopmans, A. Carolyn Olson, Ester Sztein, and Donald L. Sparks. 2015. Soil science: Soil and human security in the 21st century. *Science* 348 (6235): 1261071. https://doi.org/10.1126/science.1261071.

Bajželj, Bojana, Julian M. Allwood, and Jonathan M. Cullen. 2013. Designing climate change mitigation plans that add up. *Environmental Science & Technology* 47 (14): 8062–8069. https://doi.org/10.1021/es400399h.

Baldos, Uris, and Thomas Hertel. 2015. The role of international trade in managing food security risks from climate change. *Food Security* 7 (April). https://doi.org/10.1007/s12571-015-0435-z.

Barrett, Christopher B. 1997. Idea gaps, object gaps, and trust gaps in economic development. *Journal of Developing Areas* 31 (4): 553.

Barrett, Christopher B. (ed.). 2013. *Food security and sociopolitical stability*. Oxford: Oxford University Press.

Barrett, Christopher B. 2020. Actions now can curb food systems fallout from COVID-19. *Nature Food* 1 (6): 1–2. https://doi.org/10.1038/s43016-020-0085-.

Barrett, Christopher B. 2021. Overcoming global food security challenges through science and solidarity. *American Journal of Agricultural Economics* 103 (2): 422–447.

Barrett, Christopher B., Luc Christiaensen, Megan Sheahan, and Abebe Shimeles. 2017. On the structural transformation of rural Africa. Policy Research Working Papers. The World Bank. https://doi.org/10.1596/1813-9450-7938.

Barrett, Christopher B., and Mark A. Constas. 2014. Toward a theory of resilience for international development applications. *Proceedings of the National Academy of Sciences of the United States of America* 111 (40): 14625. https://doi.org/10.1073/pnas.1320880111.

Barrett, Christopher B., Jessica Fanzo, Mario Herrero, Daniel Mason-D'Croz, Alexander Mathys, Philip Thornton, Stephen Wood, Tim G. Benton, Shenggen Fan, Laté Lawson-Lartego, Rebecca Nelson, Jianbo Shen, and Lindiwe Majele Sibanda. 2021. COVID-19 pandemic lessons for agri-food systems innovation. *Environmental Research Letters* 16 (10): 101001. https://doi.org/10.1088/1748-9326/ac25b9.

Béné, Christophe. 2020. Resilience of local food systems and links to food security—A review of some important concepts in the context of COVID-19 and other shocks. *Food Security* 12 (4): 805–822. https://doi.org/10.1007/s12571-020-01076-1.

Benton, Tim G., Juliet A. Vickery, and Jeremy D. Wilson. 2003. Farmland biodiversity: Is habitat heterogeneity the key? *Trends in Ecology & Evolution* 18 (4): 182–188. https://doi.org/10.1016/S0169-5347(03)00011-9.

Bloom, Nicholas, Charles I. Jones, John Van Reenen, and Michael Webb. 2020. Are ideas getting harder to find? *American Economic Review* 110 (4): 1104–1144. https://doi.org/10.1257/aer.20180338.

Cattaneo, Andrea, Marco V. Sánchez, Máximo Torero, and Rob Vos. 2020. Reducing food loss and waste: Five challenges for policy and research. *Food Policy*, September, 101974. https://doi.org/10.1016/j.foodpol.2020.101974.

Chai, Yuan, Philip G. Pardey, Connie Chan-Kang, Jikun Huang, Kyuseon Lee, and Wanlu Dong. 2019. Passing the food and agricultural R&D buck? The United States and China. *Food Policy* 86 (July): 101729. https://doi.org/10.1016/j.foodpol.2019.101729.

Chaplin-Kramer, Rebecca, Megan E. O'Rourke, Eleanor J. Blitzer, and Claire Kremen. 2011. A meta-analysis of crop pest and natural enemy response to landscape complexity. *Ecology Letters* 14 (9): 922–932. https://doi.org/10.1111/j.1461-0248.2011.01642.x.

Contreras, Jorge L., Michael Eisen, Ariel Ganz, Mark Lemley, Jenny Molloy, Diane M. Peters, and Frank Tietze. 2020. Pledging intellectual property for COVID-19. *Nature Biotechnology* 38 (10): 1146–1149. https://doi.org/10.1038/s41587-020-0682-1.

Cooper, Karen A., Tom E. Quested, Helene Lanctuit, Diane Zimmermann, Namy Espinoza-Orias, and Anne Roulin. 2018. Nutrition in the bin: A nutritional and environmental assessment of food wasted in the UK. *Frontiers in Nutrition* 5 (March). https://doi.org/10.3389/fnut.2018.00019.

Dalin, Carole, Yoshihide Wada, Thomas Kastner, and Michael J. Puma. 2017. Groundwater depletion embedded in international food trade. *Nature* 543 (7647): 700–704. https://doi.org/10.1038/nature21403.

Davis, Kyle F., Jessica A. Gephart, Kyle A. Emery, Allison M. Leach, James N. Galloway, and Paolo D'Odorico. 2016. Meeting future food demand with current agricultural resources. *Global Environmental Change* 39 (July): 125–132. https://doi.org/10.1016/j.gloenvcha.2016.05.004.

Dolislager, Michael, Thomas Reardon, Aslihan Arslan, Louise Fox, Saweda Liverpool-Tasie, Christine Sauer, and David L. Tschirley. 2021. Youth and adult agrifood system employment in developing regions: Rural (peri-urban to hinterland) vs. Urban. *The Journal of Development Studies* 57 (4): 571–593. https://doi.org/10.1080/00220388.2020.1808198.

Ehrmann, Jürgen, and Karl Ritz. 2014. Plant: Soil interactions in temperate multi-cropping production systems. *Plant and Soil* 376 (1): 1–29. https://doi.org/10.1007/s11104-013-1921-8.

Evenson, Robert E., and Douglas Gollin. 2003. Assessing the impact of the Green Revolution, 1960 to 2000. *Science* 300 (5620): 758–762. https://doi.org/10.1126/science.1078710.

Fanzo, Jessica, Lawrence Haddad, Rebecca McLaren, Quinn Marshall, Claire Davis, Anna Herforth, Andrew Jones, et al. 2020. The food systems dashboard is a new tool to inform better food policy. *Nature Food* 1 (5): 243–246. https://doi.org/10.1038/s43016-020-0077-y.

FAO. 2019. *The state of food and agriculture 2019: Moving forward on food loss and waste reduction.* Rome.

Fogel, Robert William. 2004. *The escape from hunger and premature death, 1700–2100: Europe, America, and the Third World.* Cambridge Studies in Population, Economy and Society in Past Time. Cambridge: Cambridge University Press. https://doi.org/10.1017/CBO9780511817649.

Fones, Helen N., Daniel P. Bebber, Thomas M. Chaloner, William T. Kay, Gero Steinberg, and Sarah J. Gurr. 2020. Threats to global food security from emerging fungal and oomycete crop pathogens. *Nature Food* 1 (6): 332–342. https://doi.org/10.1038/s43016-020-0075-0.

Fuglie, Keith, Madhur Gautam, Aparajita Goyal, and William F. Maloney. 2020. *Harvesting prosperity: Technology and productivity growth in agriculture*. Washington, DC: World Bank. https://doi.org/10.1596/978-1-4648-1393-1.

GAVI. 2020. Gavi launches innovative financing mechanism for access to COVID-19 vaccines. June 4, 2020. https://www.gavi.org/news/media-room/gavi-launchesinnovative-financing-mechanism-access-covid-19-vaccines.

GBD 2017 Diet Collaborators. 2019. Health effects of dietary risks in 195 countries, 1990–2017: A systematic analysis for the global burden of disease study 2017. *Lancet* 393 (10184): 1958–1972. https://doi.org/10.1016/S0140-6736(19)30041-8.

Gentilini, Ugo, Mohamed Almenfi, Pamela Dale, Robert Palacios, Harish Natarajan, Guillermo Alfonso Galicia Rabadan, Yuko Okamura, John Blomquist, Miglena Abels, Gustavo Demarco, and Indhira Santos. 2020. *Social protection and jobs responses to COVID-19: A real-time review of country measures*. Washington, DC: World Bank. https://openknowledge.worldbank.org/handle/10986/33635.

Gollin, Douglas, Casper Worm Hansen, and Asger Wingender. 2018. *Two blades of grass: The impact of the Green Revolution*. No. w24744. National Bureau of Economic Research, Cambridge, MA. https://doi.org/10.3386/w24744.

Haddad, Lawrence. 2020. Viewpoint: A view on the key research issues that the CGIAR should lead on 2020–2030. *Food Policy* 91 (February): 101824. https://doi.org/10.1016/j.foodpol.2020.101824.

Herrero, Mario, and Philip Thornton. 2020. What can COVID-19 teach us about responding to climate change? *The Lancet. Planetary Health* 4 (5): e174. https://doi.org/10.1016/S2542-5196(20)30085-1.

Hoddinott, John F. 2014. *Resilience: A primer*. 2020 Conference Brief 8. May 17–19, Addis Ababa, Ethiopia. International Food Policy Research Institute (IFPRI): Washington, DC. http://ebrary.ifpri.org/cdm/ref/collection/p15738coll2/id/128159.

International Labour Organization (ILO). 2015. Agriculture: A hazardous work. Geneva: ILO. Accessed November 7, 2020. https://www.ilo.org/global/topics/safety-and-health-at-work/areasofwork/hazardous-work/WCMS_356550/lang--en/index.htm.

International Labour Organization (ILO). 2017. Working together to promote a safe and healthy working environment. Report of the International Labour Conference 2017. Geneva: ILO.

IPBES. 2019. Summary for policymakers of the global assessment report on biodiversity and ecosystem services of the Intergovernmental Science-Policy Platform on Biodiversity and Ecosystem Services. S. Díaz, J. Settele, E. S. Brondízio, H. T. Ngo, M. Guèze, J. Agard, A. Arneth, P. Balvanera, K. A. Brauman, S. H. M. Butchart, K. M. A. Chan, L. A. Garibaldi, K. Ichii, J. Liu, S. M. Subramanian, G. F. Midgley, P. Miloslavich, Z. Molnár, D. Obura, A. Pfaff, S. Polasky, A. Purvis, J. Razzaque, B. Reyers, R. Roy Chowdhury, Y. J. Shin, I. J. Visseren-Hamakers, K. J. Willis, and C. N. Zayas (eds.). IPBES secretariat, Bonn, Germany. 56 pages. https://doi.org/10.5281/zenodo.3553579.

IPCC. 2019. Climate change and land: An IPCC special report on climate change, desertification, land degradation, sustainable land management, food security, and greenhouse gas fluxes in terrestrial ecosystems. P.R. Shukla, J. Skea, E. Calvo Buendia, V. Masson-Delmotte, H.-O. Pörtner, D.C. Roberts, P. Zhai, R. Slade, S. Connors, R. van Diemen, M. Ferrat, E. Haughey, S. Luz, S. Neogi, M. Pathak, J. Petzold, J. Portugal Pereira, P. Vyas, E. Huntley, K. Kissick, M. Belkacemi, J. Malley, eds. https://www.ipcc.ch/srccl/.

Jalil, Andrew, Joshua Tasoff, and Arturo Bustamante. 2020. Eating to save the planet: Evidence from a randomized controlled trial using individual-level food purchase data. *Food Policy* 95 (August): 101950. https://doi.org/10.1016/j.foodpol.2020.101950.

Kanter, David R., Olivia Chodos, Olivia Nordland, Mallory Rutigliano, and Wilfried Winiwarter. 2020. Gaps and opportunities in nitrogen pollution policies around the world. *Nature Sustainability* 3: 956–963. https://doi.org/10.1038/s41893-020-0577-7.

Kremen, Claire, and Albie Miles. 2012. Ecosystem services in biologically diversified versus conventional farming systems: Benefits, externalities, and trade-offs. *Ecology and Society* 17 (4). https://www.jstor.org/stable/26269237.

Kremer, Michael, Jonathan Levin, and Christopher M. Snyder. 2020. Advance market commitments: Insights from theory and experience. *AEA Papers and Proceedings* 110 (May): 269–273. https://doi.org/10.1257/pandp.20201017.

Laborde, David, Marie Parent, and Carin Smaller. 2020. *Ending hunger, increasing incomes, and protecting the climate: What would it cost donors?* Washington: Ceres2030, International Institute for Sustainable Development, and International Food Policy Research Institute.

Mason-D'Croz, Daniel, Timothy B. Sulser, Keith Wiebe, Mark W. Rosegrant, Sarah K. Lowder, Alejandro Nin-Pratt, Dirk Willenbockel, et al. 2019. Agricultural investments and hunger in Africa modeling potential contributions to SDG 2—Zero hunger. *World Development* 116 (April): 38–53. https://doi.org/10.1016/j.worlddev.2018.12.006.

Maxwell, Daniel, and Peter Hailey. 2020. Towards anticipatory information systems and action: Notes on early warning and early action in East Africa. Boston: Feinstein International Center, Tufts University; Nairobi: Centre for Humanitarian Change.

Mellor, John W. 2017. *Agricultural development and economic transformation: Promoting growth with poverty reduction*. New York, NY: Palgrave Macmillan.

Munich Reinsurance. 2020. Risks posed by natural disasters. Accessed November 1, 2020. https://www.munichre.com/en/risks/natural-disasters-losses-are-trending-upwards.html#-1624621007.

Newbold, Tim, Lawrence N. Hudson, Andrew P. Arnell, Sara Contu, Adriana De Palma, Simon Ferrier, Samantha L. L. Hill, et al. 2016. Has land use pushed terrestrial biodiversity beyond the planetary boundary? A global assessment. *Science* 353 (6296): 288–291. https://doi.org/10.1126/science.aaf2201.

Ortiz-Bobea, Ariel, Toby R. Ault, Carlos M. Carrillo, Robert G. Chambers, and David B. Lobell. 2020. The historical impact of anthropogenic climate change on global agricultural productivity. ArXiv E-Prints 2007 (July): arXiv:2007.10415.

Ostrom, Elinor. 2010. Beyond markets and states: Polycentric governance of complex economic systems. *The American Economic Review* 100 (3): 641–672.

Pardey, Philip G., Connie Chan-Kang, Steven P. Dehmer, and Jason M. Beddow. 2016. Agricultural R&D is on the move. *Nature News* 537 (7620): 301. https://doi.org/10.1038/537301a.

Paulot, Fabien, and Daniel J. Jacob. 2014. Hidden cost of U.S. agricultural exports: Particulate matter from ammonia emissions. *Environmental Science & Technology* 48 (2): 903–908. https://doi.org/10.1021/es4034793.

Pelletier, Johanne, Hambulo Ngoma, Nicole M. Mason, and Christopher B. Barrett. 2020. Does smallholder maize intensification reduce deforestation? Evidence from Zambia. *Global Environmental Change* 63 (July): 102127. https://doi.org/10.1016/j.gloenvcha.2020.102127.

Pingali, Prabhu L. 2012. Green revolution: Impacts, limits, and the path ahead. *Proceedings of the National Academy of Sciences* 109 (31): 12302–12308. https://doi.org/10.1073/pnas.0912953109.

Poggio, Santiago L. 2005. Structure of weed communities occurring in monoculture and intercropping of field pea and barley. *Agriculture, Ecosystems & Environment* 109 (1): 48–58. https://doi.org/10.1016/j.agee.2005.02.019.

Polaris Market Research. 2020. *Plant-based meat market share, size, trends, industry analysis report, by source (soy, wheat, pea, and others); By product (burger patties, sausages, strips & nuggets, meatballs, others); By application (retail outlets, foodservice, E-commerce), by regions; Segment forecast, 2020 –2027*. Report PM1689. New York: Polaris Market Research.

Poore, J., and T. Nemecek. 2018. Reducing food's environmental impacts through producers and consumers. *Science* 360 (6392): 987–992. https://doi.org/10.1126/science.aaq0216.

Reardon, Thomas, and Johan Swinnen. 2020. COVID-19 and resilience innovations in food supply chains. In *IFPRI Book Chapters*, 132–136. International Food Policy Research Institute (IFPRI). https://ideas.repec.org/h/fpr/ifpric/133836.html.

Ritchie, Hannah, David S. Reay, and Peter Higgins. 2018. Beyond calories: A holistic assessment of the global food system. *Frontiers in Sustainable Food Systems* 2. https://doi.org/10.3389/fsufs.2018.00057.

Sanchez, Pedro A. 2020. Time to increase production of nutrient-rich foods. *Food Policy* 91 (February): 101843. https://doi.org/10.1016/j.foodpol.2020.101843.

Schumpeter, Joseph A. 1942. *Capitalism, socialism, and democracy*. New York: Harper and Brothers.

Siegrist, Michael, and Christina Hartmann. 2019. Impact of sustainability perception on consumption of organic meat and meat substitutes. *Appetite* 132 (January): 196–202. https://doi.org/10.1016/j.appet.2018.09.016.

Tang, Xiumei, Saiyun Luo, Zhipeng Huang, Haining Wu, Jin Wang, Guoying Shi, Liangqiong He, et al. 2019. Changes in the physicochemical properties and microbial communities of rhizospheric soil after cassava/peanut intercropping. *BioRxiv*, March, 570937. https://doi.org/10.1101/570937.

Tendall, Danielle M., Jonas Joerin, Birgit Kopainsky, Peter J. Edwards, Aimee Shreck, Quang Bao Le, P. Kruetli, Michelle Grant, and Johan Six. 2015. Food system resilience: Defining the concept. *Global Food Security* 6 (October): 17–23. https://doi.org/10.1016/j.gfs.2015.08.001.

Thurlow, James. 2020. Measuring Agricultural Transformation. Presentation to USAID, Washington, DC. Available at: https://www.slideshare.net/ifpri/aggdp-agemp-measuring-agricultural-transformation.

Tilman, David, Kenneth G. Cassman, Pamela A. Matson, Rosamond Naylor, and Stephen Polasky. 2002. Agricultural sustainability and intensive production practices. *Nature* 418 (6898): 671–677. https://doi.org/10.1038/nature01014.

Tilman, David, and Michael Clark. 2014. Global diets link environmental sustainability and human health. *Nature* 515 (7528): 518–522. https://doi.org/10.1038/nature13959.

United Nations High Commissioner for Refugees. 2020. *Figures at a Glance*. UNHCR. June 18, 2020. https://www.unhcr.org/figures-at-a-glance.html.

van Boeckel, Thomas P., Charles Brower, Marius Gilbert, Bryan T. Grenfell, Simon Asher Levin, Timothy P. Robinson, Aude Teillant, and Ramanan Laxminarayan. 2015. Global trends in antimicrobial use in food animals. *Proceedings of the National Academy of Sciences of the United States of America* 112 (18): 5649–5654. https://doi.org/10.1073/pnas.1503141112.

Van Loo, Ellen J., Vincenzina Caputo, and Jayson L. Lusk. 2020. Consumer preferences for farm-raised meat, lab-grown meat, and plant-based meat alternatives: Does information or brand matter? *Food Policy* 95 (August): 101931. https://doi.org/10.1016/j.foodpol.2020.101931.

Yi, Jing, Eva-Marie Meemken, Veronica Mazariegos-Anastassiou, Jiali Liu, Ejin Kim, Miguel I. Gómez, Patrick Canning, and Christopher B. Barrett. 2021. Post-farmgate food value chains make up most of consumer food expenditures globally. *Nature Food* 2 (6): 417–425. https://doi.org/10.1038/s43016-021-00279-9.

Zhao, Qingyun, Wu Xiong, Yizhang Xing, Yan Sun, Xingjun Lin, and Yunping Dong. 2018. Long-term coffee monoculture alters soil chemical properties and microbial communities. *Scientific Reports* 8 (1): 6116. https://doi.org/10.1038/s41598-018-24537-2.

Key External Drivers of Change to 2070

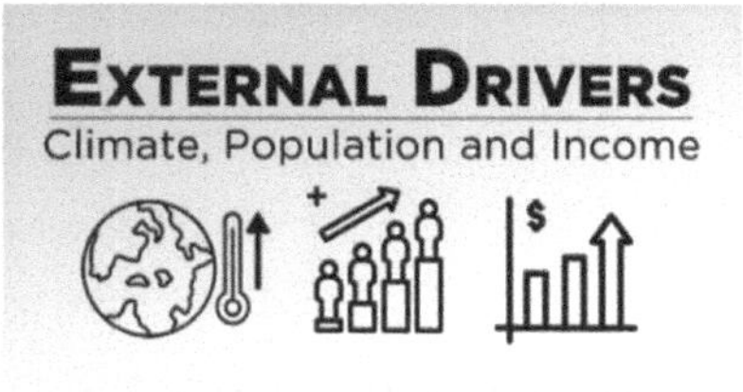

As we look 25–50 years, or more, into the future, we must also keep in mind how very different tomorrow's world will inevitably look. Three big, inevitable changes stand out, with serious implications for AFS and AVC innovations.[1]

First, **the geography of human populations will shift markedly.** The world became majority urban in 2007, and by 2050 the UN projects that 68 percent will live in cities (UN DESA 2019). This means elongated supply chains from rural breadbasket areas but also puts a premium on land-saving technologies that enable short supply chains serving significant concentrations of consumers.

The best recent projections forecast global population peaking in about 2064 at roughly 9.7 billion people (Vollset et al. 2020), an increase of roughly one-quarter of today's 7.8 billion. Even more striking, however, will be the dramatic shift in population from Europe and East Asia, where many countries' populations have already peaked or will peak this decade, to Sub-Saharan Africa, where population will continue on an upward trajectory well into the next century

[1] One might consider digitization a fourth big, inevitable driver originating largely outside of AFSs. We omit it, however, because digitization is well underway already and likely to play out largely over the coming decade or so, rather than persisting over the coming 20–50 years. As we discuss extensively below, digital technologies represent a plurality of the promising innovations being implemented already or on the near-term horizon.

C. B. Barrett et al., *Socio-Technical Innovation Bundles for Agri-Food Systems Transformation*, Sustainable Development Goals Series,
https://doi.org/10.1007/978-3-030-88802-2_3

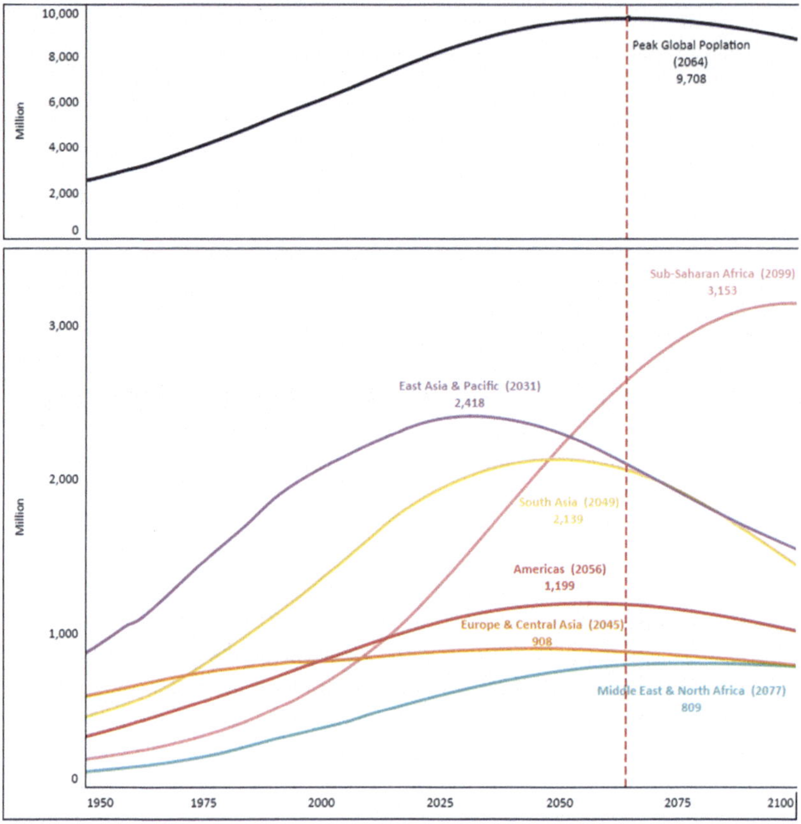

Fig. 1 Human population projections by world region, 1950–2100. Colored lines reflect regional population. Historical estimates are from 1950 to 2017, and population is projected from 2018 to 2100. Peak population and year are labeled. The peak global population is additionally reflected by the red dashed line in 2064 (*Data Source* Vollset et al. 2020)

(Fig. 1). This stems directly from the massive youth bulge in sub-Saharan Africa, where the median age is just 19 years, half of that in Europe or North America and far below even the median age of 31 in Asia and Latin America (UN Population Prospects 2019). Since more than 70 percent of food is eaten in the country in which the source commodity was grown (D'Odorico et al. 2014) and because greenhouse gas emissions typically rise with the geographic length of the supply chain, spatial patterns of population growth will compel increased attention to African AFSs and AVCs for reasons associated with all four of the design objectives.

The second major, inevitable driver of AVC changes will be income growth, especially in today's LMICs.[2] Income growth fuels increased consumer demand for food (Fukase and Martin 2020). This matters mainly because, in the market-based economies that drive AVCs today and indefinitely into the future, (often latent) consumer demand is the biggest driver by far of product and process innovation as firms adapt in search of greater market share and profits. Indeed, income growth patterns and the differential way income growth translates into food demand growth in poorer versus richer communities, along with population growth patterns mean that Africa will be the main locus of food market expansion over the rest of this century (Box 1).

Box 1: Turn attention to Africa

Researchers and policymakers increasingly recognize that in order to address the myriad challenges facing global AFSs and to meet the SDGs, we must actively attend to the needs of smallholder farmers and poor consumers in rural and traditional systems, most of them in Africa and Asia. A plurality of the world's people live in rural and traditional systems (Table 1 in Chapter 1), and they are disproportionately unlikely to be able to afford a nutritious diet (Bai et al. 2020) and suffer the world's lowest agricultural productivity (Fuglie et al. 2020). These regions most urgently need investments to co-create socio-technical bundles—the combinations of technological, policy, and institutional innovations we advocate for below—to advance HERS objectives, as efforts such as CERES2030 (https://ceres2030.org/) have demonstrated.

---

[2] Most widely-regarded (e.g., IMF, OECD) medium-to-long-run economic growth forecasts project a slowdown in world real income growth from the trend rate of 3.0–3.5 percent/year in the late 2010s, with the high-income OECD member states growing by just 1–2 percent annually, the largest middle-income economies—the so-called "BRIICS" (Brazil, Russia, India, Indonesia, China and South Africa)—decelerating from 4 to 6 percent annual growth today to just 2–3 percent/year by 2060, with growth in today's lower and lower-middle income countries, including most of Africa, overtaking the BRIICS this decade (Guillemette and Turner 2018; IMF 2020).

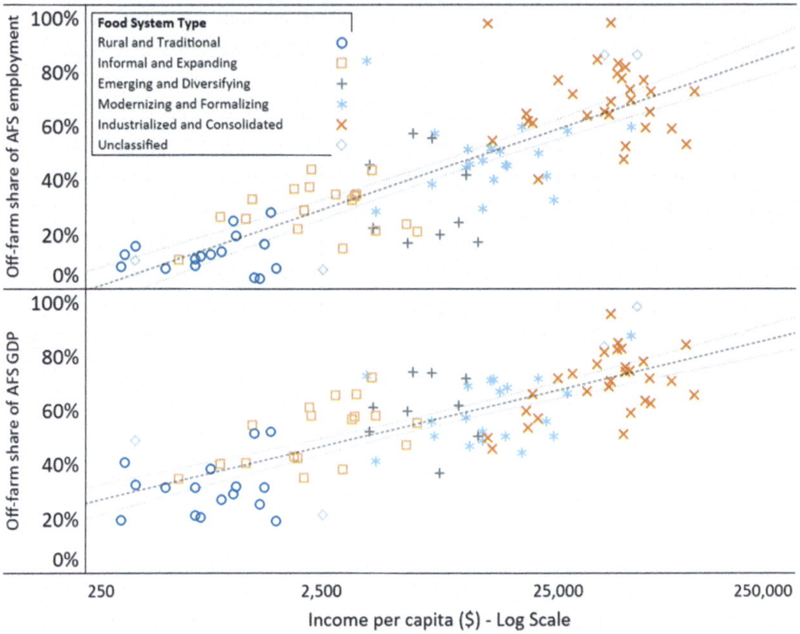

Fig. 2 Off-farm share of agri-food system GDP and employment by income per capita. Each data point reflects a country's off-farm share of employment and GDP in the AFS in the most recently reported year (generally 2015). Dashed line shows regression line relating expected share to income per capita, with 95 percent confidence bands interval reflected by dotted lines (*Data Source* Thurlow 2020)

What remains less well recognized is that growth in agri-food market opportunities arising from food demand expansion will occur overwhelmingly in Africa (Barrett 2021). In today's roughly US$8 trillion global food market, African purchases account for less than ten percent. That will change dramatically in the decades ahead. Food demand growth is largely a function of three parameters: growth in the number of people eating, the rate of per person income growth for those consumers, and the share of that income growth that converts into food demand (what economists call the "income elasticity of demand for food"). Global population growth to the end of the century will concentrate almost exclusively in Africa (Fig. 2).

The income elasticity of demand for food falls rapidly as incomes grow to, and through, the middle-income range. So the same income growth in Africa, now the world's poorest continent, will translate into much greater (double or triple) food demand expansion than in other world regions. As a result of just population growth and income elasticity of demand differences, even if Africa's per capita income growth does not continue to outpace the rest of the world, as it did 2010–2019, **a majority of global food demand growth to 2100 will occur in Africa, at least tripling the region's global market share**. Under more aggressive growth scenarios, the region could easily account for three-quarters of global food demand growth to 2100. This trend is already well underway, as the inflation-adjusted annual sales growth in Africa of food retail grocery and food service chains has far outpaced that of other world regions over the past decade (Barrett et al., in press).

Moreover, income growth does not scale food demand equally across products and processes. It mainly boosts higher-quality foods, foods that are more processed and varied, and those that are more resource-intensive (e.g., animal source proteins) as well as food prepared and eaten away from consumers' homes. The biggest demand response to income growth is non-nutritive quality attributes—appearance, convenience, safety, social status, storability, taste, and variety—as well as perceived environmental or social attributes associated with the production process (Barrett et al., in press). This naturally concentrates value addition and employment growth in the post-farmgate portions of AVCs (Thurlow et al. 2019; Yi et al. 2021), where many food product and process innovations originate, which comes through clearly when looking at the relationship between incomes and the off-farm share of both AVC employment and value addition (Fig. 2).

Third, given climate change already baked into our atmospheric systems due to GHGs of recent decades, Earth will be warmer, with changes to the start and duration of growing seasons; more severe and frequent storms, droughts, and floods; and rising sea levels, and greater irregularities (IPCC 2018). AFSs must be prepared for such conditions. Coastal production systems must adapt, logistics infrastructure must be hardened or moved, and vulnerable populations must be displaced to higher ground. Increased water scarcity, higher temperatures, and higher atmospheric CO_2 concentrations will lead to lower nutrient density in some crops and forage species; greater stress on crops, livestock, and the people who tend them; and changes in the prevalence and distribution of pests and diseases. International trade options will be increasingly important to enable rapid adaptation to pronounced regional differences in climate fluctuations (Janssens et al. 2020).

The existential threat posed by failure to get better control over both the climate and the parallel species extinction crises will compel dedicating more land to carbon sequestration in trees and soils, to habitat conservation to preserve wild species and buffer human populations against dangerous zoonoses, and to the production of renewable geothermal, solar, and wind energy to displace fossil fuels consumption. All of these functions require converting rural lands from agriculture and protecting them from industrial and residential expansion in the face of expanding cities. This will compel a partial de-agrarianization of food systems, that is, steadily reducing the land and water footprint of food production through substituting capital for land and water inputs to absorb a rising share of growing food demand (Barrett 2021).

Meanwhile, income growth will almost surely increase consumers' willingness to pay for foods' non-nutritive credence attributes[3] related to GHG emissions, environmental sustainability, animal welfare, working conditions, etc., all of which are easier to trace and certify in shorter supply chains. Increasingly urban demand and heightened consumer concerns about long supply chains in the aftermath of pandemic disruptions and trade wars will likely reinforce these patterns, as might advances in household-scale renewable energy generation and 3-D printing that make micro-scale, personalized food production increasingly viable. All of this favors emergent controlled environment agriculture, especially to produce higher-value fresh fruits and vegetables, and precision fermentation and tissue engineering methods to produce higher-value proteins to compete with traditional livestock and seafood products, as well as circular feeds designs to reduce the marine and land footprint of livestock feed production. The paths such transformations follow remain to be charted, however.

[3] Credence attributes cannot be observed by consumers after purchase and thus rely on trust, if only trust in third-party certification of the qualities for which the buyer pays a premium. Credence attributes in foods mainly relate to unobservable upstream production and exchange processes—how workers are treated, the fairness of payments to farmers, environmental impacts, even the geographic origins of the product—or to healthfulness claims. The resulting information asymmetries invite fraud in the absence of effective private or public regulation (Dulleck and Kerschbamer 2006), and the gains seem to accrue mainly to consumers and intermediaries, not to primary producers (Meemken et al. 2020).

We describe these three key drivers as external to AFSs because each process will advance regardless of the path AFSs follow. But make no mistake, AFS innovation feeds back into demographic transitions, income growth, and the climate and extinction crises. Indeed, we face real climate, environmental, health, and social dangers today and in the decades ahead in part because the past century's AFS innovations have focused so tightly on boosting agricultural productivity, especially output per unit area cultivate (i.e., yields), to the exclusion of other objectives. Nudging the coming generation of AFS innovations in better directions requires envisioning a broader set of shared objectives.

References

Bai, Yan, Robel Alemu, Steven A. Block, Derek Headey, and William A. Masters. 2020. Cost and affordability of nutritious diets at retail prices: Evidence from 177 countries. *Food Policy* (October): 101983. https://doi.org/10.1016/j.foodpol.2020.101983.

Barrett, Christopher B. 2021. Overcoming global food security challenges through science and solidarity. *American Journal of Agricultural Economics* 103 (2): 422–447.

D'Odorico, Paolo, Joel A. Carr, Francesco Laio, Luca Ridolfi, and Stefano Vandoni. 2014. Feeding humanity through global food trade. *Earth's Future* 2 (9): 458–469. https://doi.org/10.1002/2014EF000250.

Dulleck, Uwe, and Rudolf Kerschbamer. 2006. On doctors, mechanics, and computer specialists: The economics of credence goods. *Journal of Economic Literature* 44 (1): 5–42. https://doi.org/10.1257/002205106776162717.

Fuglie, Keith, Madhur Gautam, Aparajita Goyal, and William F. Maloney. 2020. *Harvesting prosperity: Technology and productivity growth in agriculture*. Washington, DC: World Bank. https://doi.org/10.1596/978-1-4648-1393-1.

Fukase, Emiko, and Will Martin. 2020. Economic growth, convergence, and world food demand and supply. *World Development* 132 (August): 104954. https://doi.org/10.1016/j.worlddev.2020.104954.

Guillemette, Yvan, and David Turner. 2018. The long view: Scenarios for the world economy to 2060. OECD Economic Policy Papers, No. 22, OECD Publishing, Paris. https://doi.org/10.1787/b4f4e03e-en.

International Monetary Fund. 2020. World Economic Outlook (October 2020). Washington: IMF. https://www.imf.org/external/datamapper/datasets/WEO.

IPCC. 2018. Global warming of 1.5 °C. An IPCC special report on the impacts of global warming of 1.5 °C above pre-industrial levels and related global greenhouse gas emission pathways, in the context of strengthening the global response to the threat of climate change, sustainable development, and efforts to eradicate poverty. Masson-Delmotte, V., P. Zhai, H.-O. Pörtner, D. Roberts, J. Skea, P.R. Shukla, A. Pirani, W. Moufouma-Okia, C. Péan, R. Pidcock, S. Connors, J.B.R. Matthews, Y. Chen, X. Zhou, M.I. Gomis, E. Lonnoy, T. Maycock, M. Tignor, and T. Waterfield, eds. Geneva: World Meteorological Organization.

Janssens, Charlotte, Petr Havlík, Tamás Krisztin, Justin Baker, Stefan Frank, Tomoko Hasegawa, David Leclère, et al. 2020. Global hunger and climate change adaptation through international trade. *Nature Climate Change* 10 (9): 829–835. https://doi.org/10.1038/s41558-020-0847-4.

Meemken, Eva-Marie, Christopher B. Barrett, Hope C. Michelson, Matin Qaim, Thomas Reardon, and Jorge Sellare. 2020. The role of sustainability standards in global agrifood supply chains. University of Copenhagen Working Paper.

Thurlow, James. 2020. Measuring Agricultural Transformation. Presentation to USAID, Washington, DC. Available at: https://www.slideshare.net/ifpri/aggdp-agemp-measuring-agricultural-transformation.

Thurlow, James, Paul Dorosh, and Ben Davis. 2019. Chapter 3—Demographic change, agriculture, and rural poverty. In *Sustainable Food and Agriculture*, ed. Clayton Campanhola and Shivaji Pandey, 31–53. Academic Press. https://doi.org/10.1016/B978-0-12-812134-4.00003-0.

United Nations Department of Economic and Social Affairs. 2019. *World urbanization prospects: The 2018 revision*, 123. https://doi.org/10.18356/b9e995fe-en.

United Nations Department of Economic and Social Affairs, Population Division. 2019. *World population prospects 2019*. https://population.un.org/wpp/Download/Standard/Population/.

Vollset, Stein Emil, Emily Goren, Chun-Wei Yuan, Jackie Cao, Amanda E. Smith, Thomas Hsiao, Catherine Bisignano, et al. 2020. Fertility, mortality, migration, and population scenarios for 195 countries and territories from 2017 to 2100: A forecasting analysis for the Global Burden of Disease Study. *The Lancet* 396 (10258): 1285–1306. https://doi.org/10.1016/S0140-6736(20)30677-2.

Yi, Jing, Eva-Marie Meemken, Veronica Mazariegos-Anastassiou, Jiali Liu, Ejin Kim, Miguel I. Gómez, Patrick Canning, and Christopher B. Barrett. 2021. Post-farmgate food value chains make up most of consumer food expenditures globally. *Nature Food* 2 (6): 417–425. https://doi.org/10.1038/s43016-021-00279-9.

Envisioning Four Design Objectives for 2045–2070

Repeated episodes throughout history remind us that **AFSs episodically undergo dramatic transformations, most of them purposeful—guided by incentives prevailing at the time—rather than purely random changes**. Typically, these changes have taken decades or centuries. A major shock, like the COVID-19 pandemic, may help spark the more rapid transformation that we desperately need. Hence the value of explicitly envisioning AFS transformation to direct the transformative power unleashed by the pandemic towards desired outcomes.

Transformations originate in either scientific or social processes, or more often a combination of the two, the sorts of socio-technical innovation bundles we emphasize in this report. All truly novel and noteworthy advances have been driven by pressing social needs, responding to economic and social incentives and harnessing the accumulated information available at the time (Arthur 2007).[1] For example, the Green Revolution's focus on dramatically expanding the supply of staple cereals and roots/tubers was directly born of concerns that insufficient supplies of dietary energy (i.e., calories) would lead to famine in the face of growing human populations (Ehrlich 1968). The Green Revolution succeeded fabulously in meeting the objective of boosting per capita calorie supplies, thereby driving

[1] Some transformative technologies originate in one sphere of society and then radically remake others. This has commonly been true of technologies developed for military purposes, such as the internet, global positioning systems, or the Haber–Bosch process.

C. B. Barrett et al., *Socio-Technical Innovation Bundles for Agri-Food Systems Transformation*, Sustainable Development Goals Series,
https://doi.org/10.1007/978-3-030-88802-2_4

down real food prices, boosting anthropometric outcomes, reducing the rate of agricultural extensification into the world's forests, and reducing infant mortality (Evenson and Gollin 2003; Gollin et al. 2018; von der Goltz et al. 2020). But the Green Revolution also had significant unintended environmental, equity, and health consequences. For the next major AFS transformation, we must design better and differently (Barrett 2021).

We therefore preface our exploration of AVC innovations that might beneficially transform the AFSs of tomorrow by first identifying the most pressing societal needs that they must address. Especially given what we know about the present state of AFSs globally, what are the key AFS design objectives for a generation or two from now, the period 2045–2070, during which we expect to reach peak human population (Vollset et al. 2020) and by which time scientific discoveries not yet made or even imagined can have matured and diffused at scale?

First, however, it is worth reminding ourselves why such design objectives matter and the remarkable transformations that can arise in response to emergent social needs. Humans began domesticating wild plants and animals roughly 12 millennia ago as semi-nomadic groups felt pressure to settle, in part to reduce episodic conflict that came from contestation of open-access resources and unplanned encounters. These early humans began to select plants based on desirable traits and to actively cultivate food crops rather than depend on hunting and gathering. The resulting domestication of wild animals and plants into the livestock breeds and crop species we know today enabled the emergence of modern civilizations.

Progress over the intervening millennia was slow and sporadic. Then the enclosure movement transformed land and labor allocation in late eighteenth– and early nineteenth–century England. Enclosure involved a sometimes-violent process of consolidating small farms and open-access lands into larger, private holdings through the exercise of economic, legal, and political power by the landed aristocracy. Enclosure is generally considered a key spark of the first modern agricultural revolution, prompting significant, sustained gains in crop productivity that were unprecedented in European history and that were generalized across the major staple crops, like barley, oats, and wheat (Fig. 1).

Gregor Mendel's mid-nineteenth-century discovery of the basic principles of heredity and use of experimental design and careful measurement laid the foundation for modern genetics and genomics but did not immediately ignite any major gains in agricultural productivity. The massive Dust Bowl droughts and Great Depression of the 1930s in the United States, however, compelled federal

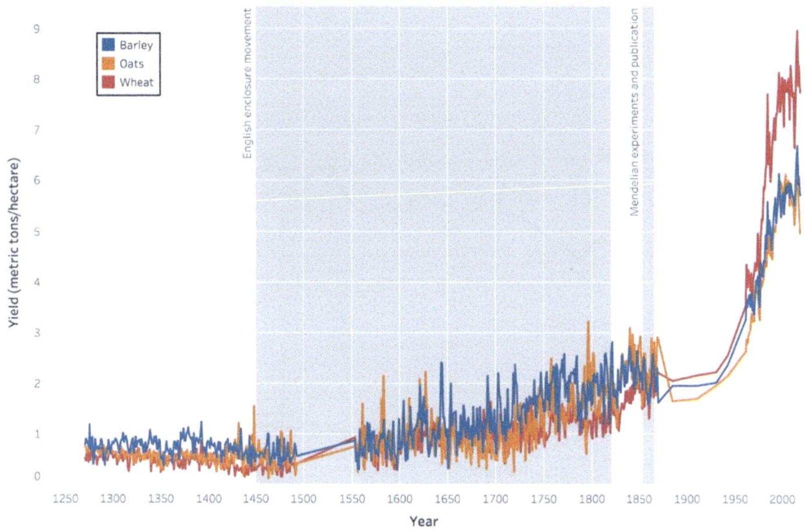

Fig. 1 Long-term cereal yields of key crops in the United Kingdom from 1270 to 2018. Wheat, barley, and oat yields are shown in metric tons per hectare (*Data sources* Our World in Data, FAOSTAT)

and state government investment in agricultural research and extension to help address mass internal migration and suffering. What followed was an extraordinary period of scientific advances in staple cereals hybridization and of labor-saving mechanization that were widely adapted and diffused, dramatically altering the agricultural productivity trajectory of the United States, with significant global spillovers that similarly transformed agriculture throughout the rest of the high-income temperate world.

Then, roughly fifty years ago, the world was staring at a "population bomb" that threatened recurring famine and mass starvation—especially in Asia and Latin America, which had not benefitted much from the temperate agriculture gains of the preceding decades (Ehrlich 1968). This ignited a Green Revolution thanks in large measure to advances in plant breeding, irrigation, and the production of inorganic nitrogenous fertilizer—and to a lesser degree, mechanization—all supported through public and philanthropic investments that ensured universal access to improved plant material, agronomic practices, and engineering designs,

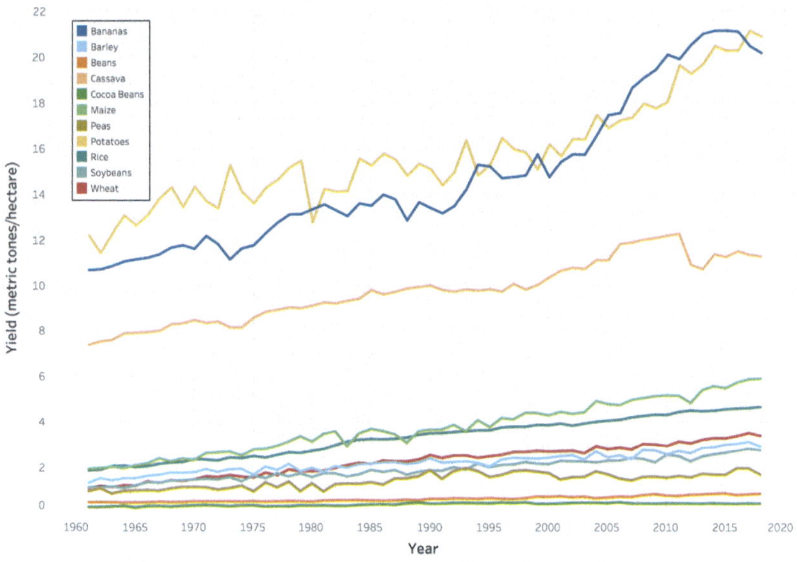

Fig. 2 Global crop yields from 1961 to 2018. Global yields for eleven staple crops are shown in metric tons per hectare (*Data source* Our World in Data based on FAO data. Note that FAO computes some crop yields based on dry grain and others based on fresh produce, inclusive of fluids)

supported by appropriate public policy and infrastructure. The resulting growth in the productivity of staple crops appropriate to a wide range of agroecologies was historically unprecedented (Fig. 2). **When faced with massive systemic, even existential challenges, our ancestors envisioned and achieved remarkable innovations that ultimately begat the AFSs we have today, for good and for ill. It is time to do so again.**

What features of the 2045–2070 world establish the design objectives for today's AVC innovators? We emphasize four essential, inter-related objectives, which we summarize with the mnemonic HERS: **h**ealthy diets, **e**quitable and sustainable livelihoods, **r**esilience to shocks and stressors, and climate and environmental **s**ustainability. The HERS objectives consolidate and build naturally on the 17 SDGs agreed to by all UN member states in 2015, especially SDGs 1 (no poverty), 2 (zero hunger), 3 (good health and well-being), 5 (gender equality), 6 (clean water and sanitation), 7 (clean and affordable energy), 8 (decent work and economic growth), 10 (reduced inequalities), 12 (responsible consumption

WHEN FACED WITH MASSIVE SYSTEMIC, EVEN EXISTENTIAL CHALLENGES, OUR ANCESTORS ENVISIONED AND ACHIEVED REMARKABLE INNOVATIONS THAT ULTIMATELY BEGAT THE AGRI-FOOD SYSTEMS WE HAVE TODAY, FOR GOOD AND FOR ILL. IT IS TIME TO DO SO AGAIN.

and production), 13 (climate action), 14 (life below water), and 15 (life on land). But these must extend far beyond the 2030 SDG target date, as few de novo innovations today stand much chance of diffusing at scale within the decade. So we take a somewhat longer-run view, beyond the 2030 horizon. We look 25–50 years into the future.

First, AFSs must meet the food security standard definition, agreed at the 1996 World Food Summit, which states, "[A]ll people, at all times, have physical and economic access to sufficient, safe, and nutritious food that meets their dietary needs and food preferences for an active and healthy life." We refer to this as the **healthy diets** objective, encompassing SDGs 2 and 3. This will require, in particular, increasing the availability of nutritious, safe, and diverse foods and ensuring adequate and continuous affordable access to, and utilization of, foods that comprise healthy diets; limiting the supply and consumption of foods that are high in refined sugars, salt, and unhealthy fats, and low in essential nutrients and bioactive compounds (e.g., carotenoids, fiber); and safeguarding foods from pathogens and contaminants (Mozaffarian 2016; Willett et al. 2019; Afshin et al. 2019). Consistent with recent HLPE (2020) recommendations, healthy diets must also respect individual food preferences, culture, and aspirations.

The second design objective is **equitable and inclusive livelihoods**, encompassing SDGs 1, 5, 8, 10, 16, and 17. Poverty is the primary cause of food insecurity throughout the world. A key driver of poverty is relatively low productivity. Most of the world's poor live in rural areas and work in AFSs. **The low economic returns to agricultural production, processing, etc., in rural and traditional systems are a key source of global inequality**. Productivity improvements accessible to the poor, therefore, have important equity implications. That is especially true for innovations that boost labor productivity among the poor because their labor power is typically their most valuable asset. They often own little land, livestock, machinery, or other forms of productive capital. People everywhere aspire to equal and inclusive opportunities but are denied basic human rights due to the accidental geography of their birth, the color of their skin, their gender, their sexual orientation, or some other identity marker irrelevant under the Universal Declaration of Human Rights. **AVCs are potentially powerful avenues to address equity and inclusion objectives**, both because they necessarily deliver life-sustaining foods and also because they provide (self- or paid) employment to well more than a billion persons worldwide.

Equity considerations require looking beyond just smallholder farmers and poor consumers to think about workers and small- and medium-sized enterprise owners throughout the AVC. Unsurprisingly, only about 2 percent of urban residents of LMICs work as farmers while about 26 percent work in the post-harvest AVC, as either enterprise owners or employees (Dolislager et al. 2021). Outside of Africa, however, even in rural areas, more people derive their livelihood primarily from AVC SMEs or farm wage labor than from their own farms, especially in Latin America (Dolislager et al. 2021).

In order to advance equity objectives, we must also cease emphasizing narrow measures of crop yields (i.e., output per unit of land cultivated) a partial productivity measure that reflects the returns to owners of land. Why? Because the poor own little or no land. The AFSs we envision for a generation or two from now will, instead, prioritize advances in total factor productivity (TFP), a measure that—when properly constructed[2]—summarizes the returns to all natural and manmade inputs, and especially in worker health and labor productivity. Greater focus on TFP will promote livable incomes for the poorest, who often possess little more than their own time.

Because adverse shocks happen, safety nets are needed so that those unable to work are assured unbroken access to healthy diets. Individuals' rights to privacy and to the personal data increasingly recordable in a digitizing world should be recognized and respected. Cultural, economic, and political life should reflect broader participation of all interested persons, decentralizing governance power while facilitating enhanced opportunities for coordination among parties.

Third, if the COVID-19 pandemic has taught us anything, it is the absolute necessity of building **resilience to shocks and stressors**.[3] As we elaborated previously, several lessons emerge from these first months of the greatest pandemic to strike the world in living memory. These lessons apply to a broad range of sources of systemic risk, not just infectious disease pandemics. Most notably, the

[2] An important criticism of TFP as typically implemented is that it ignores environmental inputs and associated externalities. As a result, TFP measures commonly overstate what is occasionally known as "total resource productivity" or "environmentally adjusted TFP," which is the real rate of advance society should seek to optimize (Fuglie et al. 2016).

[3] Stressors—often also labelled "ex-ante risk" exposure—refer to the prospect of adverse events that could strike, and that influence human behavior and well-being, but that have not yet materialized. Shocks are the ex-post realization of adverse stochastic events, whether or not they were anticipated. Therefore, stressors do not always turn into shocks, and shocks may not have been anticipated as stressors.

world faces substantial, and likely growing, risks due to climate change, violent conflicts, trade wars, etc. The likelihood of additional severe disruptions occurring within the coming generation is high.

This leads to the fourth and final design objective: **environmental and climate sustainability**, encompassing SDGs 6, 7, 11, 12, 13, 14, and 15. For TFP to work as a measure, **we must more comprehensively monitor and sustain the natural systems on which AFSs fundamentally depend**, and move away from simple partial productivity metrics, such as yield (i.e., output per unit land area cultivated), or reductionist measures of TFP that ignore nature's inputs into agri-food production. We must develop and consistently employ measures of AFS productivity: maximizing the number of people nourished healthily and sustainably while minimizing environmental and health care costs. We must also rigorously establish the thresholds beyond which agroecosystems and the climate become unlikely to recover from excessive stresses.

This will reduce the unyielding intensification pressure on scarce land and water resources. Land, at multiple scales—from field through landscape to wildland—must be spared for nature, in part to protect humankind from infectious disease. Agricultural drivers—mainly extensification of cultivated lands into forests and wetlands—are associated with more than 25 percent of all infectious diseases, and more than 50 percent of zoonotic diseases, that emerged in humans since the 1940s (Rohr et al. 2019). Anthropogenic land conversion increases the density of species that vector a broader number of dangerous viruses, as these hosts, on average, outcompete non-host species in converted lands (Gibb et al. 2020). We must value "less but better" food, with significant adoption of approaches based on agroecological principles rather than exclusive reliance on external inputs that homogenize the environment. Highly external input intensive production will and should still occur, enhanced by the principles of sustainable intensification, in areas where the net impacts are modest (e.g., avoiding areas of high intrinsic biodiversity). Sustainable intensification based on external inputs can usefully complement agroecological intensification that boosts productivity through implementation of agroecological principles at the plot, farm and landscape levels. Sustainable intensification, the rise of circular economies, and the mainstreaming of agroecological practices will have preserved, or even expanded, the necessary wild or multi-use spaces for other plant and animal species to survive and thrive, on land and below water. Air and water quality will have stabilized at healthy levels. Overall, through changing our demands for food, protecting nature from the expansion of agricultural land into new areas, and farming in more sustainable ways, we will have converted agri-food production from a net source of nearly 30 percent of climate-threatening GHG emissions to wider land use patterns that represent a GHG sink—or "zero net carbon" land use at a minimum—thereby helping mitigate the climate crisis.

References

Afshin, Ashkan, Patrick John Sur, Kairsten A. Fay, Leslie Cornaby, Giannina Ferrara, Joseph S. Salama, Erin C. Mullany, et al. 2019. Health effects of dietary risks in 195 countries, 1990–2017: A systematic analysis for the global burden of disease study 2017. *The Lancet* 393 (10184): 1958–1972. https://doi.org/10.1016/S0140-6736(19)30041-8.

Arthur, W. Brian. 2007. The structure of invention. *Research Policy* 36 (March): 274–287. https://doi.org/10.1016/j.respol.2006.11.005.

Barrett, Christopher B. 2021. Overcoming global food security challenges through science and solidarity. *American Journal of Agricultural Economics* 103 (2): 422–447.

Dolislager, Michael, Thomas Reardon, Aslihan Arslan, Louise Fox, Saweda Liverpool-Tasie, Christine Sauer, and David L. Tschirley. 2021. Youth and adult agrifood system employment in developing regions: Rural (peri-urban to hinterland) vs. urban. *The Journal of Development Studies* 57 (4): 571–593. https://doi.org/10.1080/00220388.2020.1808198.

Ehrlich, Paul R. 1968. *The population bomb*. New York: Ballantine Books.

Evenson, Robert E., and Douglas Gollin. 2003. Assessing the impact of the Green Revolution, 1960 to 2000. *Science* 300 (5620): 758–762. https://doi.org/10.1126/science.1078710.

Fuglie, Kieth, Tim Benton, Yu Sheng, Julien Hardelin, Koen Mondelaers, and David Laborde. 2016. *G20 MACS white paper: Metrics of sustainable agricultural productivity. OECD* 68.

Gibb, Rory, David W. Redding, Kai Qing Chin, Christl A. Donnelly, Tim M. Blackburn, Tim Newbold, and Kate E. Jones. 2020. Zoonotic host diversity increases in human-dominated ecosystems. *Nature* 584 (7821): 398–402. https://doi.org/10.1038/s41586-020-2562-8.

Gollin, Douglas, Casper Worm Hansen, and Asger Wingender. 2018. *Two blades of grass: The impact of the Green Revolution*. No. w24744. National Bureau of Economic Research, Cambridge, MA. https://doi.org/10.3386/w24744.

High Level Panel of Experts on Food Security and Nutrition (HLPE). 2020. *Food security and nutrition: Building a global narrative towards 2030*. Rome.

Mozaffarian, Dariush. 2016. Dietary and policy priorities for cardiovascular disease, diabetes, and obesity: A comprehensive review. *Circulation* 133 (2): 187–225. https://doi.org/10.1161/CIRCULATIONAHA.115.018585.

Rohr, Jason R., Christopher B. Barrett, David J. Civitello, Meggan E. Craft, Bryan Delius, Giulio A. DeLeo, Peter J. Hudson, et al. 2019. Emerging human infectious diseases and the links to global food production. *Nature Sustainability* 2 (6): 445–456. https://doi.org/10.1038/s41893-019-0293-3.

Vollset, Stein Emil, Emily Goren, Chun-Wei Yuan, Jackie Cao, Amanda E. Smith, Thomas Hsiao, Catherine Bisignano, et al. 2020. Fertility, mortality, migration, and population scenarios for 195 countries and territories from 2017 to 2100: A forecasting analysis for the Global Burden of Disease Study. *The Lancet* 396 (10258): 1285–1306. https://doi.org/10.1016/S0140-6736(20)30677-2.

von der Goltz, Jan, Aaditya Dar, Ram Fishman, Nathaniel D. Mueller, Prabhat Barnwal, and Gordon C. McCord. 2020. Health impacts of the Green Revolution: Evidence from 600,000 births across the developing world. *Journal of Health Economics* 74 (December): 102373. https://doi.org/10.1016/j.jhealeco.2020.102373.

Willett, Walter, Johan Rockström, Brent Loken, Marco Springmann, Tim Lang, Sonja Vermeulen, Tara Garnett, et al. 2019. Food in the Anthropocene: The EAT–Lancet Commission on healthy diets from sustainable food systems. *The Lancet* 393 (10170): 447–492. https://doi.org/10.1016/S0140-6736(18)31788-4.

Getting from Here to There

So how do we reverse the growing carbon, land, and toxic chemical footprint of contemporary AVCs; expand the nutrient-rich food supply; and induce more equitable, inclusive, healthier food environments—and thus consumption patterns—so as to navigate from today's unsustainable and precarious AVCs to a warmer, more urban, more African, and shock-prone world in which wealthier consumers place an ever-growing premium on the non-nutritive attributes of the foods they buy? Given the climate change, population and income growth, and urbanization baked into AFSs already, beneficial innovation is the only feasible pathway. And because innovation takes time, typically measurable in decades, we urgently need to accelerate innovative activity.

But **no one-size-fits-all innovations exist. Many candidate socio-technical bundles are available, but those that can work in one system may be ill-suited for others. Appropriate paths from today to tomorrow necessarily differ by context**.

The demographic, epidemiological, and nutritional transitions underway vary markedly across distinct AFSs and societies as the food environments in which people make dietary choices evolve differentially. Much of this evolution is influenced by the nutrition transition, the changes in dietary and physical activity patterns of populations primarily driven by a set of factors including increased and accelerated urbanization, globalization, and economic development in countries (Popkin et al. 2012). These changing dietary and physical activity patterns are correlated with a rise in the prevalence of overweight, obesity, and noncommunicable diseases in tandem with stymied undernutrition in LMICs (Popkin et al. 2020). Figure 1 shows how the double burden of malnutrition changes among each AFS typology (from rural and traditional to industrialized and consolidated). One clearly sees the sharp decline in child stunting prevalence as AFSs develop—

C. B. Barrett et al., *Socio-Technical Innovation Bundles for Agri-Food Systems Transformation*, Sustainable Development Goals Series,
https://doi.org/10.1007/978-3-030-88802-2_5

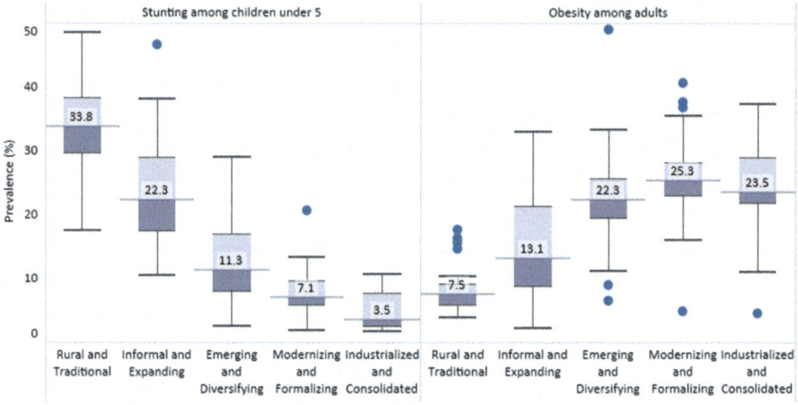

Fig. 1 Stunting and obesity by system type (*Data source* Marshall et al. 2021)

as well as the continued existence of stunting even in the most advanced systems—and the corresponding rise in the prevalence of obesity among adults.

Drewnowski and Popkin (1997) earmarked distinct patterns that cut across the nutrition transition (Fig. 2).[1] Consistent with our rural and traditional AFS typology, people in Drewnowski and Popkin's Patterns 1–3 have access to seasonally-dependent local foods, with much of their diet coming from staple grains and roots/tubers. Animal source foods are less available and affordable, and highly processed, packaged foods are sold in lower volumes, although that is changing (Baker et al. 2020). These populations are vulnerable to higher incidences of childhood wasting or stunting, high maternal and child mortality rates—often due to communicable diseases—and other factors that contribute to a shorter life expectancy (Frassetto et al. 2009; IFPRI 2015).

As economies and AFSs transition due to economic growth and urbanization, countries in Pattern 4 shift more towards those classical patterns of industrialized and consolidated AFS types. Food supply chains, markets, and environments become more varied and diverse (Barrett et al. in press). Urbanization drives demographic and technological changes so that more women enter the labor force (Seto and Ramankutty 2016). In this Pattern 4, and with transitioning and emerg-

[1] Drewnowski and Popkin's Patterns 1 and 2 (massive famines and hunter/gatherer-dominated societies) are rare in modern societies.

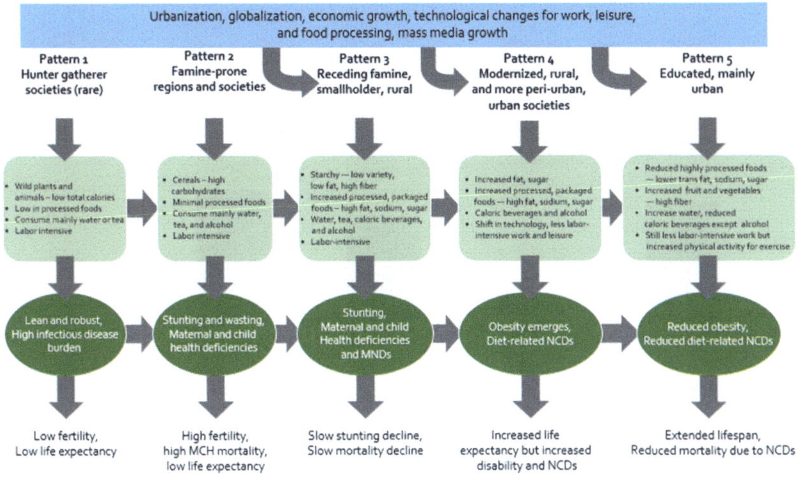

Fig. 2 The nutrition transition in five patterns (Adapted from Drewnowski and Popkin 1997)

ing AFSs, there is access to more processed and convenient foods, street food, and fast food, and more and more people consume food away from home. This is reflected partly in the strong shift towards purchasing food for home consumption in modern retail outlets, as shown in Fig. 3. Physical activity often decreases due to changes in employment type and transportation (Kearney 2010). These changes in diets and activity have important implications for the onset of overweight, obesity, and non-communicable diseases (Popkin et al. 2020). Many countries categorized as emerging and transitioning AFS types are now reeling from a double burden of malnutrition among their population (Gómez et al. 2013).

In modern or industrialized AFSs, behavioral change begins to reverse the negative tendencies of the preceding patterns, although currently this remains too rare, even in high-income settings. Figure 1 shows some suggestive evidence of modest improvements in adult obesity prevalence in industrialized and consolidated AFSs. Consumers with greater educational attainment, higher incomes, and better access to health care exhibit a higher level of concern about eating healthier and exhibit increased levels of purposeful physical activity (Popkin et al. 2012). Food acquisition also dramatically changes towards more personalized and digitized platforms. Globally, online grocery sales have grown rapidly, especially in China (Fig. 4), a trend that the COVID pandemic is expected to accelerate.

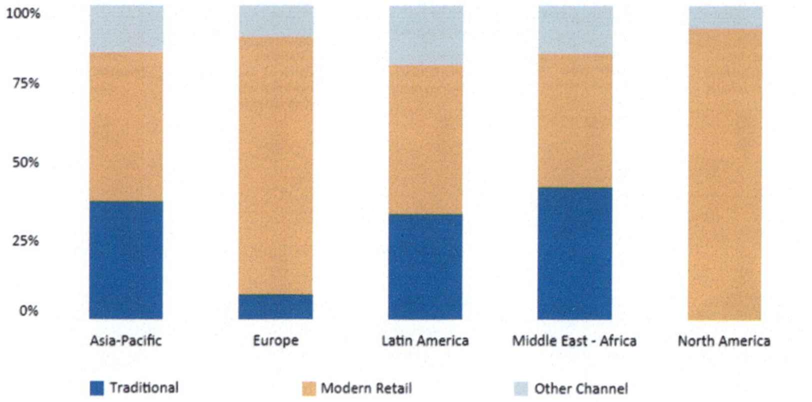

Fig. 3 Share of food purchases by type of vendor. Modern retail includes supermarkets, hypermarkets, hard discounters, and convenience stores (*Data source* Nielsen 2015)

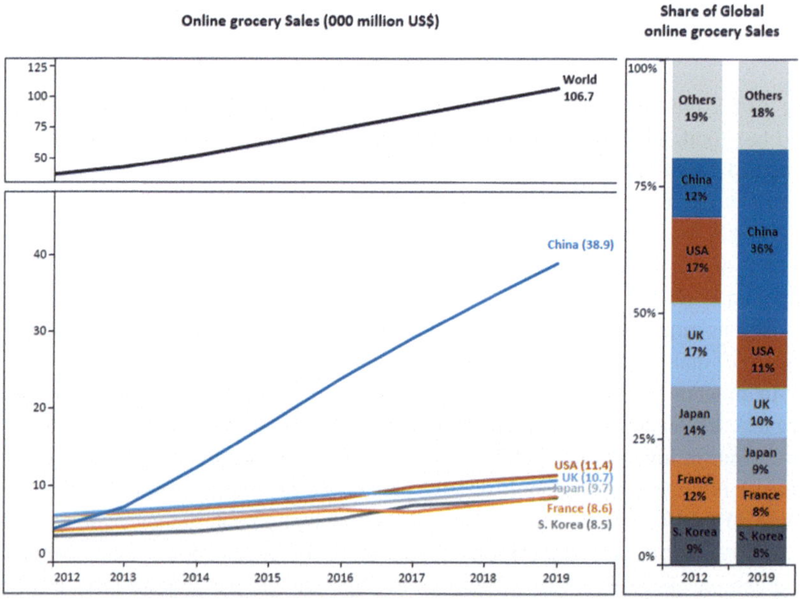

Fig. 4 Online grocery sales trends, 2012–2019 (left-hand panel) and share of global online grocery sales (right-hand panel). The data from 2012 to 2016 is historical, with 2017–2019 forecasted by Euromonitor (*Data source* Euromonitor 2017 as cited in AAFC 2017)

As we navigate change within any given AFS context, innovations do not automatically advance healthy diets, equitable and inclusive livelihoods, environmental and climate sustainability, or resilience, much less some combination of those objectives. **We must not naïvely believe that profitable innovation is inevitably favorable in all aspects relevant to society, nor that societally desirable innovations offer an attractive return on private investment**. Some scientific and social innovations may aggravate underlying dysfunction, reinforcing preexisting structures that cause, or at least aggravate, AFSs' foundational weaknesses. The discovery and upscaling of low-cost high fructose corn syrup, for example, or of some toxic chemicals were impactful, but not in especially positive ways ultimately.

Nor do discoveries with great scientific promise necessarily translate into scalable impact. The institutional environment into which innovations get introduced matter enormously to whether the resulting path leads to impact. Consider the juxtaposition of two scientific breakthroughs in rice genetics: the IR8 and IR64 varieties originated in 1966 and 1985, respectively, by the International Rice Research Institute (IRRI), and the transgenic golden rice variety revealed in 2000 that biosynthesizes beta carotene, the precursor to vitamin A. Golden rice was arguably the more impressive scientific achievement and met a pressing societal need, as reflected in the US Patent and Trademark Office recognizing it with a Patent for Humanity Award in 2015. Yet 20 years after its discovery to great fanfare, golden rice has not yet received full approval for commercial cultivation, processing, and sale in any country. By contrast, the semi-dwarf IR8 was the first "miracle rice" and the third generation IR64 became purportedly the most diffused cereal seed variety in history. The different outcomes arose less from scientific differences than from social ones. In the face of broad popular distrust of genetic engineering, and faced with a dense thicket of patents to navigate, golden rice has failed to deliver on its fanfare, while the IRRI varieties developed using conventional plant breeding methods succeeded with publicly funded R&D and extension in an environment more trusting of science, and less reliant on private funding and intellectual property protections. The juxtaposition of these advances in rice genetics underscores how innovations that advance one or more productivity, health, environmental or other objective rarely

WE MUST NOT NAÏVELY BELIEVE THAT PROFITABLE INNOVATION IS INEVITABLY FAVORABLE IN ALL ASPECTS RELEVANT TO SOCIETY, NOR THAT SOCIETAL LY DESIRABLE INNOVATIONS OFFER AN ATTRACTIVE RETURN ON PRIVATE INVESTMENT.

emerge spontaneously, given the myriad obstacles to overcome. Navigating to beneficial innovation requires proactive efforts by key actors, as well as, perhaps, a bit of good fortune.

This requires paying close attention to five key considerations simultaneously, so as to avoid linear thinking about the future. Several considerations matter to selecting appropriate innovations to advance our four design objectives. Each of these comprises a spectrum that reflects trade-offs to be considered within each specific future systems context; there is no universal right answer. The design objectives are the following:

- **Spatial extent of supply chains**: Short supply chains are often more transparent, more trusted, more valued socially, and have lower associated transport costs but may have limits on the diversity of crops available at any one time of year (Gómez and Ricketts 2013; Pradhan et al. 2020). Longer supply chains can be more efficient based on global comparative advantage—including with respect to environmental impacts (e.g., GHG emissions), given differences in transport modes—and are in some cases specific to the crop grown (e.g., coffee, cocoa, or tropical fruits that will only grow in certain regions). Localized AFSs may be more resilient to some disruptions (e.g., port and trade-related), and globalized AFSs to others (e.g., regional climate shocks). Localized AFSs may also benefit from local "ownership" (i.e., sovereignty) and thus have stronger concern for local environmental conservation, although potentially at the expense of less visible and more distant global environmental and climate objectives.
- **Scale of production**: Highly concentrated systems can sometimes offer significant efficiencies due to economies of scale and/or scope, including the ability to mobilize financing to cover the considerable fixed costs of R&D. But more concentrated systems may also pose greater systemic risks in times of crisis (as COVID-19's impact on highly concentrated meat supply chains illustrates) and be more prone to inefficient or exploitative market power. More distributed systems, on the other hand, tend to foster greater competition and perhaps also create more local ownership of problems and initiatives because AFS is integral to many communities.
- **Product diversity**: Biodiverse AFSs are commonly more resilient to myriad shocks than are ones based on fewer species. Diverse diets are also typically healthier than ones based on fewer food types, given the varied and incomplete nutrients provided by individual foods. Diversity often comes at a cost when there exist economies of scope, however. Sometimes trade-offs arise as one seeks greater diversity within AVCs.

- **Functional redundancy**: Redundancy typically increases average costs of production and distribution. Redundancy might create excess production, or wastage during storage, increasing pressure on land. But redundancy typically reduces vulnerability to systemic shocks, helps limit market power, and can promote greater diversity.
- **Internalization of externalities**: Internalizing the environmental and health costs of food so that producers bear the full costs associated with environmental degradation (e.g., biodiversity loss; impacts on air, water, and soil quality; and climate change) and public health impacts (e.g., from toxic chemicals, hazardous additives, etc.) can reduce those damages by encouraging producers to find less harmful methods. But prices will almost surely increase, which can harm poor people's access to affordable, healthy diets, unless subsidies shift to favor the affordability of nutrient-dense foodstuffs to grow and purchase.

Each of these five considerations impacts one or more of the four HERS design objectives for future AFSs: healthy diets, equitable and inclusive livelihoods, resilience, and sustainability. They help characterize the desired attributes of AFSs beyond simply minimizing the cost of calories, the primary design objective from a half century ago.

References

Agriculture and Agri-Food Canada (AAFC). 2017. *Global analysis report: E-grocery market in China*. Ottawa, CA.

Baker, Phillip, Priscila Machado, Thiago Santos, Katherine Sievert, Kathryn Backholer, Michalis Hadjikakou, Cherie Russell, et al. 2020. Ultra-processed foods and the nutrition transition: Global, regional and national trends, food systems transformations and political economy drivers. *Obesity Reviews: An Official Journal of the International Association for the Study of Obesity*, August. https://doi.org/10.1111/obr.13126.

Barrett, Christopher B., Thomas Reardon, Johan Swinnen, and David Zilberman. In press. Agri-food value chain revolutions in low-and middle-income countries. *Journal of Economic Literature*.

Drewnowski, Adam, and Barry M. Popkin. 1997. The nutrition transition: New trends in the global diet. *Nutrition Reviews* 55 (2): 31–43. https://doi.org/10.1111/j.1753-4887.1997.tb01593.x.

Frassetto, L.A., M. Schloetter, M. Mietus-Synder, R.C. Morris, and A. Sebastian. 2009. Metabolic and physiologic improvements from consuming a paleolithic, hunter-gatherer type diet. *European Journal of Clinical Nutrition* 63 (8): 947–955. https://doi.org/10.1038/ejcn.2009.4.

Gómez, Miguel I., Christopher B. Barrett, Terri Raney, Per Pinstrup-Andersen, Janice Meerman, André Croppenstedt, Brian Carisma, and Brian Thompson. 2013. Post-Green

Revolution food systems and the triple burden of malnutrition. *Food Policy* 42 (October): 129–138. https://doi.org/10.1016/j.foodpol.2013.06.009.

Gómez, Miguel I., and Katie D. Ricketts. 2013. Food value chain transformations in developing countries: Selected hypotheses on nutritional implications. *Food Policy* 42 (October): 139–150. https://doi.org/10.1016/j.foodpol.2013.06.010.

International Food Policy Research Institute (IFPRI). 2015. *Global nutrition report 2015: Actions and accountability to advance nutrition and sustainable development*. Washington, DC. https://doi.org/10.2499/9780896298835.

Kearney, John. 2010. Food consumption trends and drivers. *Philosophical Transactions of the Royal Society B: Biological Sciences* 365 (1554): 2793–2807. https://doi.org/10.1098/rstb.2010.0149.

Marshall, Quinn, Jessica Fanzo, Christopher B. Barrett, Andrew D. Jones, Anna Herforth, and Rebecca McLaren. 2021. Building a global food systems typology: A new tool for reducing complexity in food systems analysis. *Frontiers in Sustainable Food Systems* 5: 432. https://doi.org/10.3389/fsufs.2021.746512.

Nielsen. 2015. The future of grocery: E-Commerce, digital technology and changing shopping preferences around the world. https://www.nielsen.com/wp-content/uploads/sites/3/2019/04/nielsen-global-e-commerce-new-retail-report-april-2015.pdf.

Popkin, Barry M., Linda S. Adair, and Shu Wen Ng. 2012. Now and then: The global nutrition transition: The pandemic of obesity in developing countries. *Nutrition Reviews* 70 (1): 3–21. https://doi.org/10.1111/j.1753-4887.2011.00456.x.

Popkin, Barry M., Camila Corvalan, and Laurence M. Grummer-Strawn. 2020. Dynamics of the double burden of malnutrition and the changing nutrition reality. *The Lancet* 395 (10217): 65–74. https://doi.org/10.1016/S0140-6736(19)32497-3.

Pradhan, Prajal, Steffen Kriewald, Luís Costa, Diego Rybski, Tim G. Benton, Günther Fischer, and Jürgen P. Kropp. 2020. Urban food systems: How regionalization can contribute to climate change mitigation. *Environmental Science & Technology* 54 (17): 10551–10560. https://doi.org/10.1021/acs.est.0c02739.

Seto, Karen C., and Navin Ramankutty. 2016. Hidden linkages between urbanization and food systems. *Science* 352 (6288): 943–945. https://doi.org/10.1126/science.aaf7439.

A Profuse Pipeline of Promising Options

Because AFSs are diverse, dynamic, and evolve continuously, they require massive continuous investment to enable ongoing discovery and adaptation merely to prevent backsliding. Major advances in science and engineering are necessary to realize the vision of equitable, inclusive, sustainable AFSs, but they are not sufficient, as human institutions and behaviors fundamentally mediate the translation of scientific discoveries into the sorts of impacts the world needs from its AFSs over the coming decades.

Too many candidate innovations exist for us to enumerate in great detail here.[1] And surely many more innovations not presently (widely) anticipated will

[1] The collaborative online innovations portal we compiled in collaboration with Project Disrupt, goes into much greater detail. One can explore the Innovative Food System Solution portal at https://ifssportal.nutritionconnect.org/.

C. B. Barrett et al., *Socio-Technical Innovation Bundles for Agri-Food Systems Transformation*, Sustainable Development Goals Series,
https://doi.org/10.1007/978-3-030-88802-2_6

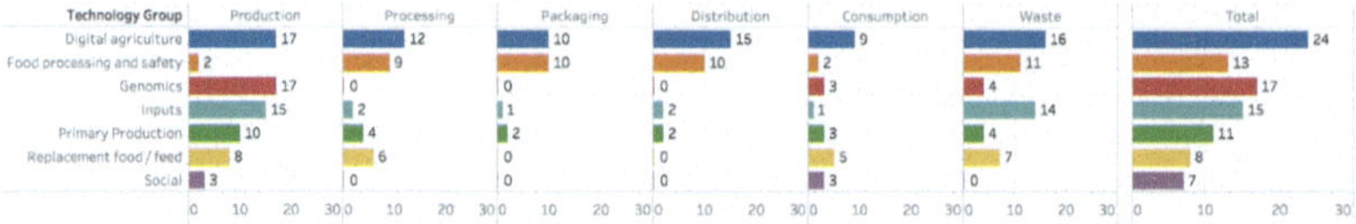

Fig. 1 Promising emergent technologies span the AVC (Adapted from Herrero et al. 2020)

emerge serendipitously or strategically in the years ahead. We know, however, that a tremendous range of options exist, spanning the full range of AVCs, from input suppliers, through retailers and food service firms (Herrero et al. 2020). Figure 1 shows that amongst the domains of cellular and digital agriculture, food processing and safety, health, and resource use efficiency, many potentially disruptive technologies span the whole AVC. Digital innovations are especially cross-cutting and numerous. From applications of molecular printing, artificial intelligence, robotics, and the Internet of Things, all the way to biodegradable coatings, new drying methods, personalized food, and the circular economy, all could have meaningful impacts through AVCs. The likely impacts and suitability of any of these inevitably vary among contexts. We take comfort in knowing that **an ever-growing pipeline of innovations could be applied in different combinations to solve particular local problems**. This diversity of innovations already under development or in various stages of adaptation and diffusion demonstrates **that multiple entry points exist to transform AFSs** (Box 1: Prioritizing Interventions).

Scientific breakthroughs generally take a significant time to incubate and evolve into more than prototypes for wider application. For example, variants of controlled environment agriculture, 3D printing of foods using AVC waste materials, and drones have each been under development for decades already. Private R&D investments typically take 5–15 years to generate discernible payoffs and public and philanthropic R&D funding, which is typically targeted at more basic scientific questions, averages 15–25 years to peak return (Chavas et al. 1997).

Nevertheless, the pipeline is healthy and ever expanding. The innovation pipeline is also increasingly well supported by private venture capital that finances an entrepreneurial ecosystem of start-up companies in the agri-food space, perhaps especially for digital agri-food technologies (Graff et al. 2020).

The innovations we studied exhibit a wide range of technological readiness, from innovations already being implemented in multiple locations and sub-sectors to ones that remain targets for basic science research (Fig. 2). A portfolio approach is necessary when thinking about the array of options. Some innovations could have very specific niches, others could be implemented in large domains. Some could have small impacts, others very large ones, as well as a variety of costs and time for implementation. Virtually all will require some—but differing types—of adaptation to suit specific AFS contexts.

Note, too, that **the most promising innovations are not solely, or even primarily, scientific breakthroughs or engineering advances**. Many key "change accelerators"—to use Herrero et al.'s (2020) term—will be sociocultural, policy, or institutional innovations because "transformation is also a deeply political process with winners and losers, which involves choices, consensus as well as compromise about new directions and pathways. Powerful players within agri-food systems have strong incentives to maintain the status quo and their current markets share" (Herrero et al. 2020, p. 267). At the same time, lucrative opportunities exist for those players that choose to help lead AFS transformation, aligning their purpose (and fortunes) to broader societal interests. Some novel organizational forms (e.g., B corporations) directly embrace such opportunities, but even some that follow more traditional organizational forms (e.g., publicly-traded, multinational corporations) are exhibiting real leadership with the expectation that this will bring both social and financial reward.

Significant differences of perspectives exist among experts concerning the potential and desirability of scientific/technological innovations now emergent. In the process of creative destruction of innovation, inevitably some people see progress, while others justifiably worry about prospective harms. There are sure to be unrealized aspirations, unanticipated consequences, predictable problems, and unforeseen obstacles, just as there will be major breakthroughs, some of them scientific, some of them sociocultural or political. Pluralism, intellectual curiosity, and healthy skepticism are paramount in advancing beneficial innovation. Innovation within AVCs is therefore far more than merely a scientific or commercial or technological matter. **Innovation is a sociopolitical phenomenon requiring ongoing consultation and monitoring** if we are to navigate successfully towards the SDGs and the longer-run design objectives of AVCs that promote healthy diets, equitable and inclusive livelihoods, environmental and climate sustainability, and resilience to shocks and stressors.

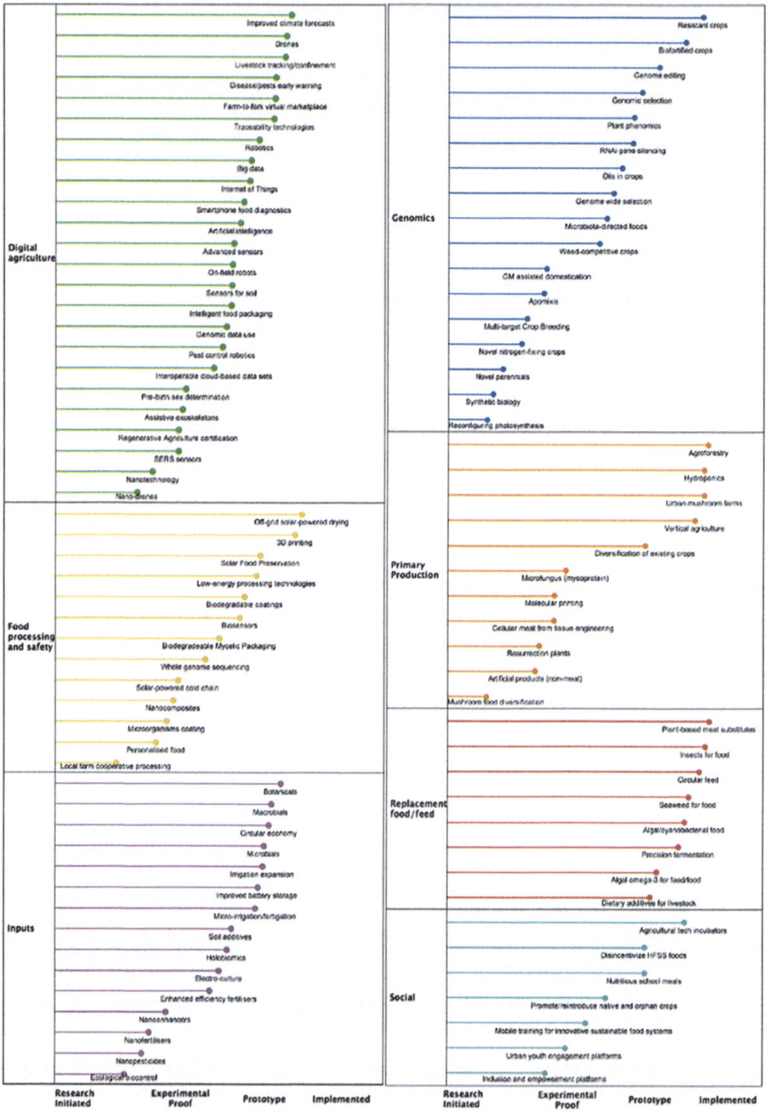

Fig. 2 Technological readiness of future agri-food systems technologies (Adapted and expanded from Herrero et al. 2020)

Given these various pressures confronting AFSs now and in the future, what AVC innovations are most likely to induce healthier diets; more sustainable and resilient production, processing, and distribution systems; and most equitable and inclusive livelihoods? Our panel identified scores of options that appear especially promising in different contexts, in distinct AVC segments, and at different time horizons. We start with four cross-cutting innovation spaces—digital, finance, social protection, and civic engagement—before moving to innovations more anchored in specific AVC stages from farm- and fisheries-based primary production through supply chain intermediaries (e.g., manufacturers, processors, and retailers), to consumer-level health and nutrition innovations. Because of the deep heterogeneity among AFSs and the considerable uncertainty around expected impacts—especially among innovations in the early stages of technological readiness—we make no attempt to rank among these. Moreover, we do not claim to offer a comprehensive listing, given the rapid pace of new discovery in the agri-food space. The sheer volume of promising innovations illustrates, however, that technological options are abundant. The key constraints relate to adapting and scaling innovations to achieve intended impacts, satisfactorily addressing unintended impacts, and setting the right incentives for beneficial innovations to emerge at sufficient pace and scale to transition AFSs towards HERS outcomes while we have time to skirt calamity.

Box 1: Prioritizing interventions for climate-smart agri-food systems*

Technologies will have different impacts on the attainment of different AFS-related SDGs. This is crucial, as different countries—or regions within countries—have achieved different levels of progress towards the different goals. Different countries might, therefore, preferentially focus on making more progress on some goals than others.

As an example from a climate-smart lens, a Delphi panel of experienced agricultural, food, and global change scientists from around the world ranked the technology list from Herrero et al. (2020) on readiness, adoption potential, and potential impact. Several technologies seem to balance readiness, adoption potential, and impacts. The top ten ranked innovations include four technologies relating to replacement food and feed for humans, livestock, and fish: plant-based substitutes, insects, microalgae and cyanobacteria, and seaweed. Driven in large part by concerns about the harmful net environmental impacts of the livestock sub-sector and how income and population growth might magnify that damage, many promising efforts are underway attempting either to meet the growing demand for animal products by providing alternative protein sources that do not rely on livestock or to reduce livestock's impacts on land via animal nutrient sources alternative to traditional feed crops.

Other top-ten technologies include improved climate forecasts and pest/disease early warning that rely on digital advances; circular economy approaches for reusing, recycling, and repurposing waste resources to boost food production while creating new local business opportunities; and vertical farming in confined spaces with no soil or natural light, another way to decouple food production from the land.

*This box draws on material from Herrero et al. (2021).

Digital Innovations

The ecosystem of digital agriculture has exploded in recent years, with the emergence of myriad agri-tech and downstream ventures across the Global South and North. The broader digital ecosystem can be envisaged as a "digital agri-stack." The foundation is made up of the macro-level enabling environment—including connectivity, human capital, and critical data infrastructure—functionality that enables system interoperability, and supporting policies. The second layer is the ecosystem of data and content. At farm level this might consist, for example, of soil and water maps; remote sensing weather data from drones, satellites, and other platforms; farmer profiles; data on animal and plant genetics; local market price and plant disease information alerts; and data from a wide range of sensors. Finally, the products and services that make use of these first two layers comprise the top layer. The various tools and applications can include distinct and bundled services spanning agricultural extension, finance, government support programs, and various advisory services. Figure 3 represents this digital agri-stack concept within the specific application domain of small farmers in rural and traditional AFSs. COVID-19 has increased the value of digital linkages in the food system, enabling people to connect to markets and production, and allowing processing and distribution operations to continue, while reducing the human contact rate of conventional approaches.

Digital technologies have penetrated even into rural and traditional contexts with notable speed, with game-changing innovations often coming from LMICs. In large part, this is due to the advancement and ubiquity of a key digital infrastructure component: the mobile phone (increasingly the smartphone) and wireless connectivity (especially as 4G becomes ubiquitous). Mobile phones have enabled people in Africa and South Asia to leapfrog over generations. But there remain significant inequities. Mehrabi et al. (2020) found that 74–80 percent of farms of larger than 200 hectares had high-speed 3G or 4G connectivity com-

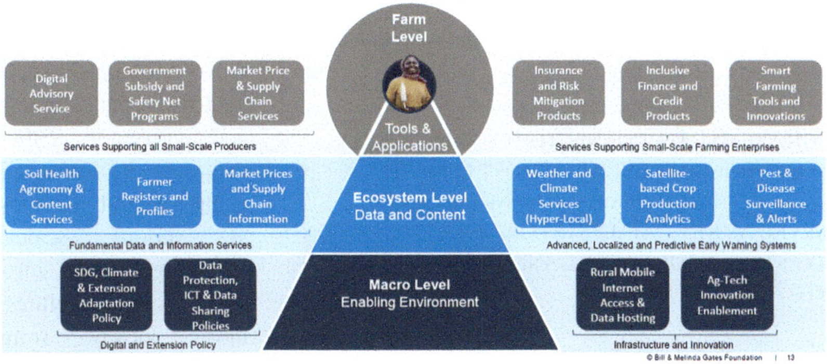

Fig. 3 The digital agri-stack (*Source* Bill and Melinda Gates Foundation, Digital Farmer Services Strategy)

pared to just 24–37 percent of farmers cultivating less than one hectare. Farms with the lowest yields and where farmers face the most climate-related shocks and food insecurity had even less digital connectivity. Data costs in Africa remain high with less than 40 percent of farming households having access to the Internet.

At the same time, digital technologies like mobile banking (e.g., mPESA in Kenya), satellite-based risk management tools (e.g., index-based livestock insurance in Ethiopia and Kenya: https://www.drylandinnovations.com/), interactive agricultural extension (e.g., Digital Green: https://www.digitalgreen.org/), and equipment sharing apps (e.g., Hello Tractor in Nigeria: https://hellotractor.com/) obviate market failures that previously constrained poor, rural populations. These digital solutions typically augment existing in-person networks, like Digital Green's partnership with government extension agents and Hello Tractor's engagement with local entrepreneurs. Those local service providers are essential; digital providers can extend their reach but not compensate for their absence or inefficiencies (Jensen and Barrett 2016). Service availability gaps thereby limit the gains to closing the digital divide.

High-speed data connectivity and smartphones in even the most remote rural communities have nonetheless served as key catalysts for new investments of capital and talent into AFSs. Affordable data pricing and design features that enable neophyte accessibility (e.g., voice recognition) are other key elements of the stack that enable the full range of stakeholders to take advantage of digital advances. Policies and regulations are also needed to create and protect trust and allow the system to continue growing and evolving. Key to this is establishing and enforcing standards that protect data privacy.

The enabling environment of connectivity and confidence facilitates the development and exploitation of critical datasets that then empower the performance of many apps. Some data is collected from users, raising issues of the rights associated with data suppliers and aggregators. Other data can be collected using spectral methods at various scales, including remote sensing from satellites, more locally from drones, and by end users or agents with hand-held devices. Remotely sensed data are increasingly available at ever higher resolution, tagged with metadata to enable their utilization. Analysis of spectral data by machine learning enables

HIGH-SPEED DATA CONNECTIVITY AND SMARTPHONES IN EVEN THE MOST REMOTE RURAL COMMUNITIES HAVE NONETHELESS SERVED AS KEY CATALYSTS FOR NEW INVESTMENTS OF CAPITAL AND TALENT INTO AFSs.

inferences that can provide users with useful information, such as rapid and low-cost estimates of key indicators of crop identity and health. Though some of these tools are expensive for small-scale producers today, their costs are reducing quickly and the increasing interoperability of sensors, data sets, and cloud-based computational tools enables the sorts of productivity-sustainability synergies that were originally envisioned from—but never fully delivered by—precision agriculture technologies introduced into industrialized AFSs starting in the 1990s (Basso and Antle 2020). The African Cassava Agronomy Initiative (http://acai-project.org/), for example, has brought together data systems, digital interfaces, and analytics to support farmers across a range of channels from mobile apps to paper.

Farmers need such hybrid apps to access reliable, accessible data on soil and weather, for instance. Also, spatially explicit datasets combined with machine learning, for example, can be used to make inferences that are useful to farmers to guide decisions on planting and crop management. Critical datasets for a healthy food system go beyond that to include data that inform actors and actions that manage food quality and safety; track and tap labor markets; provide credit and insurance markets; map nutritional status; monitor sources of pollution, etc. Data pipelines need to not only source raw data, but crucially, to analyze and transform it so that it can be interpreted and acted upon. These require investment to maintain but benefit everyone, including the private sector.

Important innovation is taking place in areas that are critical to farmer livelihoods, including farm advisory services, digitally linking market actors more efficiently, and supporting more efficient product aggregation among farmers. Some extension-based apps enable more precise, efficient, and effective use of seeds and fertilizers, while others provide disease diagnostic services for animals, plants, and people (e.g., PlantWise, https://www.plantwise.org/). Other applications support peer-to-peer (P2P) learning networks supporting entrepreneurs and other service providers who serve as intermediaries in the technical space. Digital capacities can link producers into farmer research networks that collectively build the evidence base (Nelson et al. 2019). Innovative Farmers (https://www.innovativefarmers.org/), for example, is a P2P innovation network that facilitates the building of farmer field groups that have a common challenge to address. Each group is paired with a trained researcher, who guides the members through experimental design and evaluation (e.g., evaluating non-pesticide pest control when neonicotinoids were banned in the EU).

Apps also link farmers, intermediaries, and markets, letting farmers understand and navigate pricing, and enabling farmers and intermediaries to more efficiently aggregate products. This has ignited demand for new infrastructure, like digitally enabled warehouses for logistics providers (e.g., Arya, https://aryacma.co.in, which operates over 1.6 million metric tons of digitally enabled agricultural commodity storage across rural India).

Digital technologies are also helping consumers trace the origins of the foods they consume, stimulating new potential behaviors in the marketplace, as consumers discover an array of innovative products and producers. In India, Stellapps (http://www.stellapps.com) is developing a digital layer traversing the country's massive dairy industry, providing dairy cooperatives and private dairy processors full transparency across the supply chain. Blockchain technology can enhance the visibility of producers, and farm-to-fork virtual marketplaces can further enhance traceability. For example, FishCoin (www.fishcoin.co) is supporting and rewarding supply chain actors who share data that enable full traceability across complex global supply chains.

Digital technology is also simply making it easier for consumers and manufacturers to access what they want, which is increasingly a more direct connection to farms and farmers. In Nigeria, Agriple (www.agriple.com) is connecting farmers with buyers to improve transparency, efficiency, and waste reduction. In China, ecommerce platform Pinduoduo (http://en.pinduoduo.com) has helped more farmers sell online as they turn ecommerce into a social experience that helps consumers learn about farming practices and get group discounts. In India, Ninjacart (http://ninjacart.com) is disintermediating fresh produce value chains, linking farmers directly with shopkeepers and consumers. Similar efforts are also underway across the online grocery sector in India, with direct farmer sourcing for fruits and vegetables initiated by multiple players vying for dominance, including BigBasket, Amazon, and Zomato. Such digital products and services can reduce scale advantages, broadening access to certification processes and high-return markets, leveling playing fields for smaller farms and downstream AVC enterprises, and facilitating more direct P2P and business-to-consumer exchange so as to reduce concentrated power in AVCs. These initiatives have scaled with the liberalization of telecommunications and agricultural policies, and have also piggybacked on government investments in roads, storage, and logistics services.

In industrialized and consolidated AFSs, the public sector publishes key data such as weather and public goods. These data can play an important role in policy design and implementation (Capalbo et al. 2017). Private sector digital tools and technologies have built upon those data and are changing how crops are planted, monitored, harvested, and consumed. The agritech startup ecosystem has driven much of this change, with large agribusinesses acquiring some of the most successful new ventures to ensure their traditional business models are not left behind. Notably, Monsanto (now Bayer) acquired The Climate Corporation (http://climate.com) in 2013, providing them with a strong digital backbone and products like Climate FieldView, which delivers field-level insights to farmers. In 2017, DuPont (now Corteva) acquired Granular (https://granular.ag) to ensure they, likewise, would have farm management software and precision agriculture solutions to provide their farmer customers. The same year also saw John Deere acquire Blue River Technology (http://www.bluerivertechnology.com), a robotics startup that had developed "see-and-spray" technologies for weed control, leveraging computer vision and artificial intelligence.

More recently, hyper-funded farmer platforms like Farmers Business Network (https://www.fbn.com) and Indigo (https://www.indigoag.com) have sought to redesign more fundamental aspects of the agricultural economy in the modern advanced, industrialized AFS. Farmers Business Network is trying to break the dominance of the Big 4 (Bayer, BASF, Corteva, and Syngenta) in seeds and crop protection chemicals, helping farmers to make data-driven decisions regarding inputs, while also providing access to crop insurance, commodity brokerage, financing, and other services. Indigo, on the other hand, started out developing innovative microbial products but is now focusing more on building a post-harvest marketplace for grain sales, commodity transportation solutions, and a carbon sequestration platform. Both players are well capitalized, having raised a total of US$571 million and US$1.17 billion from investors (including US$250 million and US$360 million in August 2020 alone) for Farmers Business Network and Indigo, respectively.

Other notable innovations enjoying accelerating farmer adoption in more industrialized AFSs in recent years include field-based Internet-of-Things systems integrating sensors and agronomic algorithms (e.g., CropX, https://www.cropx.com); precision agriculture platforms to optimize farm equipment (e.g., Solinftec, https://solinftec.com); and farm robotics, where startups have initially focused on weeding use cases (e.g., Naio Technologies, https://www.naio-technologies.com, and ecoRobotix, www.ecorobotix.com) but are now beginning to tackle harvesting (e.g., Advanced Farm Technologies, https://www.advanced.farm, as well as Root AI, https://www.root-ai.com), and dairy parlor management.

Further downstream, we find agritech startups working to improve post-harvest supply chains, often focused on creating better market linkages between farmers and buyers. In the American heartland, Bushel (https://bushelpowered.com) has set out to "facilitate clear and simple business

EVEN WITH AN ENABLING ENVIRONMENT (CONNECTIVITY AND DATA) AND A RANGE OF APPS, THE PERFORMANCE OF THE DIGITAL AGRISTACK CAN UNDERPERFORM IF THE ECOSYSTEM OF DATA AND INSIGHTS BECOMES FRAGMENTED.

between grain companies and growers" through a mobile application. In California (and other horticultural regions), Full Harvest (https://www.fullharvest.com) has created a digital business-to-business (B2B) marketplace to help growers sell ugly and surplus fresh produce to food processors and other potential buyers. Imperfect Foods (https://www.imperfectfoods.com) has a similar mission of fighting food waste but works across grocery categories and delivers "rescued foods" directly to customers' homes. Finally, GrubMarket (https://www.grubmarket.com) has carved a niche for itself as a farm-sourced version of Instacart, offering food products at wholesale prices to both B2B clients and consumers.

Even with an enabling environment (connectivity and data) and a range of apps, the performance of the digital agristack can underperform if the ecosystem of data and insights becomes fragmented. The adoption of a common technology platform can help AFS actors converge and continue to enhance system performance. An *enabling* technology platform for any ecosystem has several components. For instance, different tools help us find data (*data discovery*), translate it into a common form (*data transformation*) and make it easy to share (*data transfer*). These services are neutral to content but required to make it easy for individuals and organizations to safely exchange or combine data. FarmStack, currently under development by Digital Green, is developing a decentralized architecture comprised of P2P connectors specifically addressing this last service. P2P connectors ease the data exchange process between partners. Of course, as data are sensitive and valuable assets, data owners want to protect them, and farmers need to have the ability to monetize their own farm and farmer profile information which other third parties can leverage to build their own applications. Therefore, FarmStack is also developing and codifying usage policies that will ease and automate this over time, accelerating the exchange of public and proprietary data to drive collective impact as well as inform policy-makers and research.

One example of a new data layer that could foster the emergence of innovative new products and services throughout AVCs is the Rockefeller Foundation-supported Periodic Table of Food Initiative (PTFI), a global effort to create a distributed network of labs using standardized methods to populate and continuously update a database of the full biochemical composition and function of food using the latest mass spectrometry technologies and bioinformatics. The current scientific understanding of food covers, at most, 150 of foods' biochemical components, typically summarized as sample averages in conventional nutrient composition databases. A food system that supports human and planetary health, however, requires rigorously collated data covering the full range of the tens of thousands of biochemical molecules in food that mediate the relationships between food, diet, health, nutrition, and the environment. PTFI can enable interoperability of data and democratize the analysis of food with the development of low-cost kits, standards, methods, cloud-based analytical tools, and a self-sustaining, broadly accessible database—the Periodic Table of Food—that will include the quantitative and qualitative analysis of foods. These data can better equip AVC actors to personalize diets and promote health for individuals based on specific needs, development stage, age, health status, and other factors, as well as improve agricultural systems for increased environmental sustainability and resilience to various biotic and abiotic shocks.

An interoperable platform of services can speed up the potential for new insights, services, and products. As already mentioned, a broad range of apps have already emerged, from physical products (e.g., robotics and smart sensors) to human networks (e.g., crowdsourcing production insights or plant disease surveillance) to conceptual (e.g., a predictive analytics app that helps farmers decide what to plant based on likely weather and market conditions). Future applications will also solve problems where we cannot yet see clear connections at the intersection of agriculture and health, financial services, and human rights, easily integrating data and producing unforeseen relationships and solutions. The barriers to more novel digital innovations in AFS primarily arise from consumer and farmer acceptance. For example, a price premium has emerged for older-model farm machinery in secondhand markets in the US as farmers seek simpler equipment that is cheaper and easier to maintain and to safeguard their privacy from equipment dealers and service providers.

Unintended consequences will almost surely arise from digital innovations, as with any new technologies. For example, digital marketplaces that help farmers sell ugly or surplus produce may deprive food banks and pantries of an important source of healthy foods to provide their patrons, or they could siphon demand from other growers, thereby depressing prices small farmers receive. Individual-specific data can enable retail or food-service marketing campaigns more effectively targeted to manipulate consumers' weaknesses or allow agro-input dealers to bundle inputs and services to extract greater profit by exploiting detailed knowledge of farmers' behaviors. If too many digital innovations get locked up in patents, it could slow advances and make IP-protected new technologies unaffordable for lower-income subpopulations. And high-tech solutions can never fully overcome natural inferential limits to generating precise, field- or farm-level information on soils and other key variables that influence farmer decision-making (Schut and Giller 2020).

Digital innovations is one space not struggling to secure adequate private investment right now. Access to adequate finance does not, however, characterize most of the AFS innovation space.

Innovative Financing

Product and process innovation inevitably requires significant up-front investment. CERES2030 (https://ceres2030.org/) recently estimated that donors need to more than double their annual contributions targeted towards food security and nutrition objectives, increasing them by an additional US$14 billion to 2030. In addition, developing-country governments will need to commit a further US$19 billion each year, just to meet three of the five targets under SDG 2 (zero hunger), with two-thirds of this additional public spending focused on Africa (Laborde et al. 2020). This assumes that government and donor spending will crowd-in an extra US$52 billion in private investment annually. This US$85 billion/year estimate is almost surely a lower bound on the scale of financing needed to transition AFSs in the developing world to meet the broader HERS objectives—not just three of five targets under SDG 2 to 2030—much less the financing needed globally.

Inducing sufficient investment in AFS innovation will require innovations in finance. The resources exist. Global assets under management at the end of 2019 stood at US$89 trillion (Heredia et al. 2020). And with interest rates at historic lows, investors actively seek promising new investment opportunities. But most capital is allocated by private investors, who presently lack incentives to address environmental or public health externalities, or to attend to needs in low income countries where limited purchasing power and weak institutional and governance frameworks depress commercial potential. To effectively exploit food-demand growth over the coming generation—especially in sub-Saharan Africa where the bulk of additional demand will occur—AVC innovations must address pervasive climate, environmental, health, and social justice spillovers in order to ensure long-term, sustainable returns. Some recent innovations and a growing pool of capital searching for aligned opportunities show promise for helping foster accelerated AFS R&D finance and for growing investment in AVC innovators committed to advancing HERS-consistent AVC transformation.

Historically, much critical basic science funding came from governments and philanthropies. That was true of the US agricultural revolution of the 1930s–1950s and of the Green Revolution in Asia and Latin America in the 1960s–1980s. But outside of a few middle-income countries (Brazil, China, and

India) that have invested heavily in agricultural R&D due to the strategic importance of the sector to their economies, public investment in agri-food R&D slowed dramatically over the 1990s and 2000s (Pardey et al. 2016). Some of this decline was due to complacency in the wake of Green Revolution successes, and there has been renewed interest on the part of some donors since the 2008–2012 global food-price spikes. Most recently, in 2019, multiple bilateral and foundation donors committed to a major expansion of funding for the CGIAR—the main network of multinational agricultural research institutions—as part of structural reforms to that global agricultural research organization. But public and philanthropic investment remains woefully insufficient to meet the yawning innovation needs of AVCs.

Reallocation of current government farm subsidies offer an obvious source for public finance for AFS innovations. Subsidy programs in most Organization for Economic Cooperation and Development (OECD) countries and in China largely impede, rather than advance, necessary innovations towards more sustainable, resilient, inclusive, and equitable AFSs (OECD 2020; Searchinger et al. 2020). **Only one-eighth of total government support of agriculture presently goes to R&D, inspection and control systems, and rural infrastructure—the things that promote beneficial innovation**—as compared to three-quarters provided as financial transfers to individual producers, mostly in a distributionally regressive manner that reinforces inequality (OECD 2020). One centerpiece of a strategy to mobilize private finance involves fixing the distorted incentives created by government agriculture subsidies that implicitly promote investment in practices and products that generate serious environmental and health spillovers. Agricultural subsidy reform is politically fraught everywhere but essential to get market signals right to induce investors to divest from unsustainable and unhealthy enterprises. The high-level Financing Nature report emphasizes "harmful subsidy reform" as its top recommendation for mobilizing finance to stem the looming biodiversity/extinction crisis (Deutz et al. 2020).

The largest and growing share of agri-food R&D investment comes from private firms (Pardey et al. 2016) (e.g., by machinery, fertilizer, and agrochemical manufacturers; seed companies; food processors and manufacturers; retailers; and food or third-party logistics enterprises). Their commercial objectives can dovetail nicely with broader societal interests in circumstances where prospective beneficiaries are able, and willing, to pay for improved products and processes, and where effective regulatory oversight or appropriate tax policies limit any negative externalities that arise from the innovation. Just as private agri-food R&D has increasingly dominated the innovation landscape over the past generation (Pardey et al. 2016), so has public awareness that modern AVC innovations commonly lead to uncorrected climate, environmental, health, and social justice externalities and fail to address the needs of the poor, who rarely present a lucrative market to investors. Simply mobilizing more capital under current financial market designs seems an unlikely path to success.

Innovation in private investment will be necessary to advance beneficial AVC innovation and finance the widespread adoption of innovations. One modest, but important, development is the rise of institutional investors with a longer-term view on returns. Whether driven by social and environmental concerns, rising concern about the downside risk of stranded assets, diminishing returns to more conventional assets, or some other motive, private investors, pensions, and others with decades-long returns horizons are increasingly investing in regenerative agriculture, sustainable forestry and fisheries, green bonds, etc. For example, as of 2019 there was more than US$320 billion under management in assets focused on regenerative agriculture in the US alone (Electris et al. 2019). Equally exciting is the emergence of a robust and growing conservation finance movement, mobilized by groups such as the Coalition for Private Investment in Conservation. Conservation finance is developing new financial instruments that are attracting private investment in financially attractive conservation investments (Deutz et al. 2020).

Across the globe the momentum is building around Environment, Social, and Governance (ESG) investing, which offers a set of recognized criteria that value-based investors can use to deploy capital for sustainable, long-term financial gains that align their principles with those of their shareholders (Boffo and Patalano 2020). While ESG rating methodologies and standards continue to be refined, the broader impact-investing market—of which ESG is only a part—has risen sharply over the last decade, now encompassing at least US$715 billion in assets under management (GIIN 2020). The impact-investing market is widely expected to grow further as evidence mounts on the positive relationship between ESG investment and corporate financial performance (Friede et al. 2015). Many ESG funds are allocated by specialized asset managers (i.e., Paris-based Livelihoods Funds) that have emerged to pool resources from private companies—including massive AVC corporations—for investments in sustainable agriculture in smallholder farming communities around the world.

While shifting investor preferences create new opportunities and unlock additional capital, thus far this remains a modest share—ten percent or less—of global private assets under management and a fraction of the resources required to trigger necessary AVC transformation. The first challenge to overcome is a geographic one. **Although international financial markets increasingly integrate economies around the globe, investment capital remains anchored to**

high-income countries by home country bias. Agroecosystems exhibit huge heterogeneity, however. Place-specific R&D is therefore essential as are localized AVC innovators and enterprises to drive adoption of HERS products and services at scale.

ALTHOUGH INTERNATIONAL FINANCIAL MARKETS INCREASINGLY INTEGRATE ECONOMIES AROUND THE GLOBE, INVESTMENT CAPITAL REMAINS ANCHORED TO HIGH-INCOME COUNTRIES BY HOME COUNTRY BIAS.

Given that most growth in food demand will take place in Africa (Box 1 in Chapter 3)—where agri-food productivity lags and environmental, healthy diet, equity, and inclusion concerns are legion—that is the continent most in need of AVC investment capital. Governments and international donors can help, but catalyzing private investment is essential and currently woefully insufficient. The simple reason is that African markets are widely perceived as less lucrative and higher risk than are high-income markets, so private investment flows lag far behind where they need to be. The bulk of private agricultural R&D investment worldwide is undertaken by a small number of massive firms; less than two dozen firms accounted for more than 70 percent of global private agricultural R&D from 1990 to 2014 (Fuglie 2016). Private agricultural R&D in developing countries accounts for only two percent of global R&D investment in the sector globally (Fuglie 2016).

We can only increase private investment in agri-food innovation in developing countries by adjusting investor incentives and designing enabling environments to promote and direct investor appetite. Innovative ideas with considerable potential are already being successfully employed. For example, the growing Green, Social, and Sustainability Bonds movement seeks to coordinate major international financial institutions and other significant players in global financial markets to support a framework intended to catalyze ESG investments. The International Capital Market Association (ICMA) has been mandated to develop a set of guidelines and principles for bond market issuers, to ensure that participants deploy and manage raised capital to facilitate and support green, socially-conscious, and sustainable investing.[2]

[2] A systematic mapping of Green, Social and Sustainability Bonds financing to which ICMA seeks to contribute is available at https://www.icmagroup.org/green-social-and-sustainability-bonds/.

Capital markets have already responded positively. Moody's projects US$400 billion in global green bond issues in 2020,[3] continuing a sharp growth trend of approximate market doubling every 2–3 years. This is likely to trigger additional inflows into underserved markets, such as Africa. A landmark agreement signed in September 2019 between Japan's Government Pension Investment Fund—the world's largest—and the African Development Bank (AfDB), supporting inclusive and sustainable growth in Africa, led to an oversubscribed US$3 billion AfDB Social Bond that was the largest USD denominated social bond transaction in capital markets when issued in March 2020.

A bottleneck to unleashing the full potential of sustainable financing remains the formalization of coherent, transparent, and standardized definitions of, and ratings for, various classes of projects. Generalized endorsement of taxonomies that can underpin regulations and generate the capacities, instruments, and reporting frameworks to appropriately steer capital flows are important. At the global level, the International Platform on Sustainable Finance (IPSF)—whose growing membership currently represents roughly half of the world's population and half of global GDP, and which also emits half of the planet's GHGs—is promoting information disclosure standards, policy frameworks, and a global governance architecture consistent with stimulating private investment and steering capital towards ESG objectives (IPSF 2020). This complements the Harmonized Framework for Impact Reporting for ESG investments endorsed by eleven leading international financial institutions.[4] In general, ESG instruments—and particularly the "Governance" component—are expected to have a significant, positive stimulus impact on private financing for emerging and developing economies where risks arising from uncertain information quality, unclear institutional frameworks, and weak governance have limited investment to date.

Even as these regulatory frameworks develop within the international public arena to promote private sector investment, businesses are also developing their own certification processes to signal their values and position themselves to attract both aligned investors and consumers. The emergence of certified B corporations (B corps)—a pro-social business form that puts environmental and social performance on par with financial performance—has shown promise as a way to internalize the true social costs and benefits of an enterprise's activities.

[3] https://www.moodys.com/research/Moodys-Green-social-and-sustainability-bond-issuance-to-jump-24--PBC_1212910.

[4] Details available at https://www.icmagroup.org/assets/documents/Regulatory/Green-Bonds/Handbook-Harmonized-Framework-for-Impact-Reporting-220520.pdf.

B corps are hybrid enterprises that legally commit to third-party environmental and social audits conducted by B Lab, a US-based non-profit organization. In many ways, B corps epitomize social entrepreneurship. But since their emergence as a distinct organizational form in the 2000s, only about 3,500 companies in 74 countries have adopted this form—including significant ones in the AVC space, such as Ben & Jerry's, Cabot, Danone North America, and Klean Kanteen—and these have not yet had a major impact on AVC innovation or investment patterns.[5] Greater creativity and innovation remain necessary to mobilize finance for agri-food R&D.

Another class of promising innovations to unlock financing—particularly suited to cutting-edge R&D—comes from advanced market commitments (AMCs), wherein governments or donors guarantee a sufficient scale of remunerative purchases of any innovation that meets pre-specified impact criteria. AMCs aim to induce private investment and ensure subsequent access to the technology by low-income users. AMCs have been used successfully for pneumococcal vaccine (Kremer et al. 2020). Many lessons remain to be learned about AMC design, but the pneumococcal experience thus far is estimated to have resulted in 700,000 lives saved at a highly favorable cost/benefit ratio (Kremer et al. 2020). Other innovation incentives (prizes, contests, etc.) likewise show promise (Wagner 2011) and are currently being implemented in the AgResults prize competitions operated by the World Bank and in the Food System Vision Prize sponsored by the Rockefeller Foundation.

A SIMPLE CHANGE TO PATENT LAWS IN HIGH-INCOME COUNTRIES COULD SIGNIFICANTLY BOOST INCENTIVES TO AGRI-FOOD R&D FOR THE GLOBAL SOUTH.

Another approach is to modify intellectual property rights. Patents offer inventors a government-sanctioned monopoly in a novel and useful discovery for a period of time in exchange for public release of all the technical details necessary to replicate the innovation. Patents' lucrative prospective returns lure large investments. Currently, however, the meager prospective monopoly returns to innovations in orphan crops in low-income countries, simple irrigation technologies suitable for the Global South, or crop drying technologies for small-scale farmers,

[5] Figures as of end-August 2020, per B Lab's web site (https://bcorporation.net/). See Cao et al. (2017) for a history of B corps, and Moroz et al. (2018) for a series of studies on their strategies and impacts.

etc., result in private under-investment in R&D for low-income markets. **A simple change to patent laws in high-income countries could significantly boost incentives to agri-food R&D for the Global South**. The idea is reasonably straightforward (Barrett 2020). In its patent application, an inventor would volunteer to dedicate its patent to the public—that is, forfeit its right to deny licensing to third parties, thereby relinquishing its monopoly supply right—in exchange for an extension of an alternate, existing patent on a non-essential product, meaning one not needed to safeguard life or essential liberties. For example, a firm with a highly profitable patent on treatments for male hair restoration[6] might profitably extend that patent for several years if, and only if, it were to develop a non-toxic means of eradicating a pest like fall armyworm or diseases like East Coast fever or black sigatoka that afflict low-income tropical agroecosystems, for which there is likely little commercial profit but great humanitarian benefit (Barrett 2020). The essence of the idea is to induce investment in socially beneficial innovations by firms that can extract monopoly rents from high-income consumers' demand for luxury products and services[7] but that could not easily recoup investments from the new discovery's primary intended beneficiaries.

The other challenge to mobilizing finance for beneficial innovation surrounds how to monetize spillover effects on the environment and third parties, including future generations. Partly, such concerns motivate public and philanthropic investment. But regulation and tax policy can also reduce the returns on activities that generate negative externalities. Combined with subsidies to those actions that generate positive externalities, the regulation and tax policies together can correct market failures and induce greater pro-social private R&D, as well. Hence the value of taxes on unhealthy highly processed foods and emissions of GHG and other pollutants, and subsidies for on-farm conservation, investment in renewable energy fixed capital, and employment of workers from marginalized subpopulations.

Some HERS objectives can be advanced through regulatory requirements on banks, insurers, and publicly traded corporations to disclose environmental and

---

[6] One could imagine many such examples of lucrative, nonessential, patent-protected discoveries, including smart phone apps or digital file compression methods, performance-enhancing devices for recreational goods (e.g., skateboards, ski bindings), pet clothing, etc.

[7] The target would be patented product with high income elasticity but low price elasticity of demand—that is, goods demanded mainly by high-income populations that are sufficiently price insensitive such that firms can extract significant monopoly rents.

social impacts of investments as a fiduciary duty to investors and society. The stronger disclosure frameworks being promoted to induce high standards of ESG performance, and the certification and reporting instruments being established to increase the confidence of impact investors can, likewise, increase the efficiency and targeting of tax and subsidy incentives. When combined, for example, with markets to facilitate emissions trading and improved technologies for monitoring and verifying nutrient fluxes—and enhanced screening, verification, and tracking of investments and innovation impacts facilitated by digital innovation and artificial intelligence—the potential to monetize the provision of environmental services can be a powerful inducement to increase beneficial investment in HERS-consistent R&D and enterprise.

Innovative Social Protection Instruments

Transformation inevitably brings dislocation. **Facilitating inclusive transformation requires effective social protection instruments to protect those who stand to lose out from creative destruction**. Otherwise, the human costs of innovation become grave and can prompt damaging backlash and associated sociopolitical instability (Barrett 2013). We witness this today in the rise of nationalist, populist political movements worldwide at a time of significant technological change that has concentrated gains among a privileged few while destabilizing many.

The Nobel Laureate Amartya Sen famously wrote, "Starvation is the characteristic of some people not *having* enough food to eat. It is not the characteristic of there *being* not enough food to eat" (1981, p. 1, emphasis in original). If we are to advance equity, inclusion, and healthy diets objectives, then demand-side innovations must accompany the supply-side ones that usually attract most of the attention in discussion of agri-food systems. Perhaps paramount among these are enhanced coverage and effectiveness of social protection instruments.

Social protection instruments aim to protect individuals from unnecessary human suffering of any sort, including diet-related ill health and extreme poverty. The idea behind social protection is to catch people who fall into hardship and assist them until they are able to sustain themselves again, thereby both preventing descents into poverty traps in which deprivation becomes self-reinforcing and encouraging productive risk-taking by instilling confidence that one will be supported in the event of misfortune (Barrett et al. 2019). Social protection instruments represent the main demand-side innovations essential to AFS transformation.

Social protection programs of various types have expanded dramatically over the past generation or so—and especially during the COVID-19 pandemic

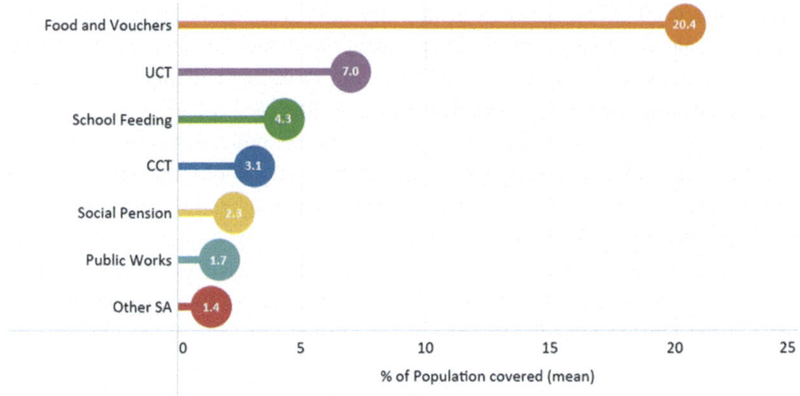

Fig. 4 Social protection program coverage among 108 low- and middle-income countries (*Data source* Alderman et al. 2017) (CCT = conditional cash transfer; SA = social assistance; UCT = unconditional cash transfer)

(Gentilini et al. 2020a)—with different purposes and impacts. But as shown in Fig. 4, food assistance programs remain the dominant mode in LMICs, covering at least 20 percent of the population—1.5 billion people—in 108 countries, more than double the total covered by conditional and unconditional cash transfers and nearly five times the population covered by school feeding programs (Alderman et al. 2017). Bastagli et al. (2016) found that at least 130 LMICs have at least one unconditional cash transfer, and about 63 have at least one conditional cash transfer. The International Labour Organization, nonetheless, estimates that only 45 percent of the global population is effectively covered by at least one social protection benefit and that developing countries alone need to invest an extra US$1.2 trillion to close their annual social protection financing gap (ILO 2020).

Myriad forms of social protection exist. Some provide a substitute for income and may include cash and in-kind transfer programs to directly boost incomes through policies (e.g. universal basic income, employment guarantee schemes, labor-intensive public works programs). Others reduce the cost of essential goods (e.g., food subsidies, vouchers, food stamps). Still others provide mechanisms to ensure access to essential public services (e.g., school scholarships, fee waivers for health care services, universal rural broadband access). Some of the most widespread and politically popular social protection programs are food assistance programs that aim to directly enhance food access (Alderman et al. 2017)—for example, through the provision of public works employment paid in food, increased purchasing power (through the provision of food stamps, coupons, or vouchers), and food-based relief interventions (through the direct provision of food to households or individuals). Some are carefully targeted in an attempt to focus coverage on specific subpopulations only (e.g., girls, orphans, and vulnerable children; the elderly; refugees; school children). Some programs only confer benefits conditional on participants engaging in specific, mandated behaviors (e.g., keeping children enrolled in school, contributing labor effort to public works programs, etc.)

Figure 5 depicts how different social protection programs fit together, depending on the targeting, mode, and conditionality of transfer. Many countries operate multiple such programs (e.g., the public distribution system and the national rural employment guarantee scheme in India, two pillars of that nation's broader welfare system) as illustrated in Fig. 5.

Given the variety of social protection programs already in use across most countries in the world, why is innovation needed in this space? Three main issues need attention. Each of these involves some combination of technical advances based on science and engineering. Mainly, however, innovations in social protection require social support and political will to invest in equitable, inclusive outcomes. **Particular attention is needed to overcome longstanding, systemic discriminatory access** on the basis of ethnicity, gender, race, religion, etc., that exists—but manifests differently—in virtually all countries. Active efforts are commonly needed to address underlying inequities through targeting and differentiation in benefits.

First, abundant evidence suggests that food-related social protection programs improve beneficiaries' lives, particularly for households that suffer from a food security shock (Behrman and Hoddinott 2005; Behrman and Skoufias 2006; Alderman et al. 2017; Hidrobo et al. 2018). But demonstrated food-related gains have been concentrated mainly on caloric acquisition and food expenditures (Hidrobo et al. 2018). The impacts on dietary diversity and quality, perhaps especially among children, remains mixed (de Groot et al. 2017). For example, a meta-analysis evaluating 15 different safety-net programs found that the impacts on child growth were insignificant overall, but at the same time, demonstrated impacts on growth in Brazil, Colombia, Ecuador, Mexico, South Africa, and Sri Lanka (Manley et al. 2012; de Groot et al. 2017).

PARTICULAR ATTENTION IS NEEDED TO OVERCOME LONGSTANDING, SYSTEMIC DISCRIMINATORY ACCESS ON THE BASIS OF ETHNICITY, GENDER, RACE, RELIGION, ETC., THAT EXISTS—BUT MANIFESTS DIFFERENTLY—IN VIRTUALLY ALL COUNTRIES.

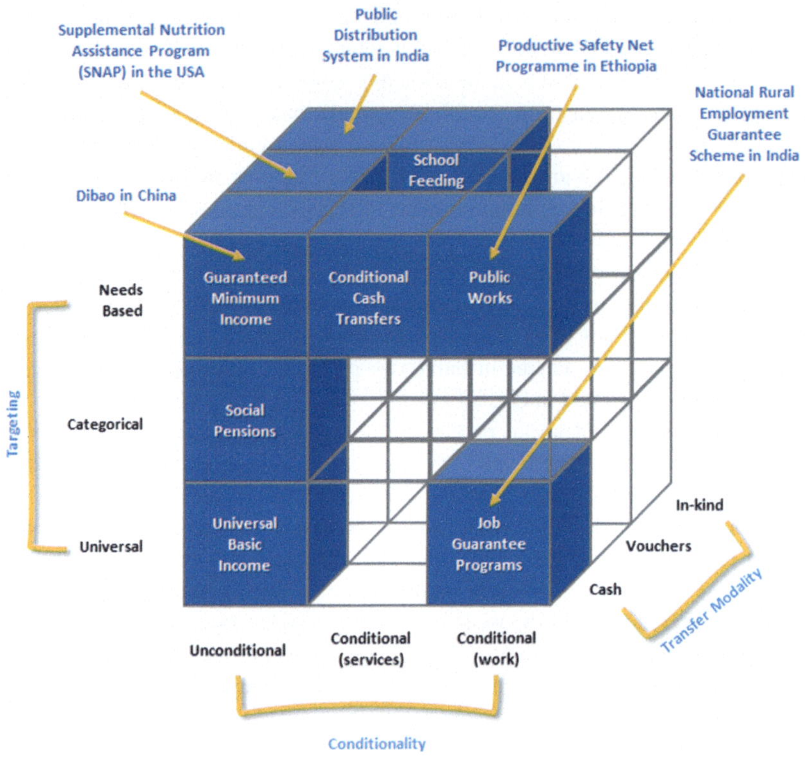

Fig. 5 How different social protection measures fit together (Adapted from Gentilini et al. 2020b)

Conditional cash transfer and food transfer or subsidy programs in Mexico, Egypt, and the US improve elements of diet quality, food insecurity, poverty, and undernutrition outcomes (Hawkes et al. 2020). Indeed, some social protection programs inadvertently increase diet-related risk of obesity and non-communicable diseases (Kronebusch and Damon 2019; Hawkes et al. 2020). So the first direction of necessary innovation is to redesign food-related social protection programs around healthy diets objectives, rather than merely avoiding hunger and undernourishment. This could occur through improved food reformulation, food assistance programs more restricted to nutrient-rich (rather than calorie-dense) foods, or other methods. Mainly, it requires political will and a change in mindset.

Second, although social protection programs' coverage appears widespread, huge numbers of people slip through the safety nets, especially marginalized populations in remote areas. Errors of inappropriate exclusion can be reduced through several directions of innovations. First, as climate, disease, trade, and other risks loom larger, we need improved early-warning systems to trigger prompt expansion of social protection programs to meet changing and growing needs. Some of this can be accomplished through advances in remotely sensed or crowd-sourced data collection, combined with advanced analytics to generate accurate, near–real time indicators of evolving needs and food supply conditions (Jean et al. 2016; Fanzo et al. 2020; Lobell et al. 2020; Porciello et al. 2020; Yeh et al. 2020).

Systems could be developed to improve the adaptive management of response form—as sometimes households most need cash to meet varied food, health, and other needs, and at other times (e.g., in hyperinflationary environments) in-kind food transfers can provide essential protection against food-price spikes. Continued advances in reliable, low-cost, secure transmission of mobile cash and vouchers can also accelerate response to safeguard healthy diets for vulnerable populations, especially in conflict-affected areas where delivery is costly and dangerous, and rapid response is of greatest humanitarian importance. Finally, diverse social protection programs remain remarkably unintegrated. Digital and other technologies can more effectively network large-scale programs (e.g., those provided by national governments), with more informal, local, and/or private food assistance programs (e.g., through food banks and pantries), automating enrollment and distributing resource demands more effectively.

Third, **the increasing digitization of social protection programs poses real risks to individual privacy and dignity**. Biometric methods can help reduce fraud to ensure prudent use of scarce public resources. But when only marginalized populations are subjected to facial, fingerprint, or other recognition tools, or if it is only beneficiaries whose personal data are made available to private vendors and service providers that may prey on underinformed consumers, programs intended to help the vulnerable can become tools of exploitation, discrimination, and disadvantage.

Innovations in Civic Engagement and Policy

The same risks of dislocation that necessitate innovations in social protection equally demand advances in civic engagement and the crafting and conduct of public policy. AFSs are both highly complex and evolving very rapidly. Technical innovations too often tend to be "pushed" (i.e., originating from R&D and effective marketing of new discoveries) rather than "pulled" (i.e., from citizens asking for new ways of doing things). And the silo-ed organization of innovation and governance ecosystems too often lead to the emergence of new products and practices developed from a reductionist, rather than a systemic, perspective to meet narrow commercial, political, or scientific aims rather than assessing synergies and tradeoffs more broadly to anticipate whether innovations will likely prove "system positive" in terms of their total impacts through positive and negative feedback pathways. The result is too often unintended, but rather predictable, adverse consequences or unfulfilled promises. The fact that our AFSs are growing ever more complex and, at the same time, the future is becoming more uncertain requires new ways of thinking to achieve "system positivity," and underpinning knowledge becomes increasingly key (though, as discussed below, knowledge—like technology—is usually necessary but rarely sufficient to engender change). As we emphasize further in the next section on socio-technical bundles, institutional innovation in civic engagement seems especially important now: innovations in engagement for AVC actors, both upstream and downstream; for policy development; and for public support (particularly through farm subsidy reform). **We urgently need both technological and institutional advances that counter concentrated commercial and political power**, in order to ensure authentically beneficial and inclusive innovations.

Innovations in upstream AVC actor engagement grow increasingly feasible in a rapidly digitizing and globalizing world. As discussed above, digital innovations are increasingly empowering farmers and food producers to innovate within their own circumstances—to adapt to challenges, adopt new opportunities, and/or harness the wisdom of crowds. Digital technologies increasingly enable connectivity that can facilitate greater inclusion in agri-food innovation and in shared governance, so as to accelerate and broaden impacts. Robust engagement is limited largely by connectivity, which reinforces the need for universal rural broadband.

Just as innovations are needed to network upstream producers (i.e., farmers, fishers, herders, etc.) more effectively, so, too, is broader citizen engagement in AVC governance, a pressing issue. In some senses the world is awash with data, yet the data we need to make informed decisions is often difficult to find. This frequently occurs because data are proprietary, because there is no regulated

WE URGENTLY NEED BOTH TECHNOLOGICAL AND INSTITUTIONAL ADVANCES THAT COUNTER CONCENTRATED COMMERCIAL AND POLITICAL POWER, IN ORDER TO ENSURE AUTHENTICALLY BENEFICIAL AND INCLUSIVE INNOVATIONS.

requirement for data transparency, or because data curation services are proprietary or absent. Artificial intelligence has made significant advances in collating, curating, and identifying associative relationships between different data streams but remains an imperfect science. Given advances in sensor and distributed ledger (e.g., blockchain) technologies to verify key details of production, transformation, and distribution processes and to store those data in nonmanipulable forms, it is increasingly feasible for citizens to access—should they so want—detailed data on how the food they consume is produced, transported, and processed, as well as evidence of its safety, the contractual terms of its production and sale, etc. Social media is also driving greater transparency by shining a spotlight on issues of consumer concern (e.g., unfair labor practices, unsustainable or unsafe production processes, etc.). Individual-level accessibility to such data, bringing down the energy and financial costs of its production, are high priorities in order to enhance civic engagement in AFS governance.

Policy is often set and managed by government departments—ministries or public agencies that are highly silo-ed, with little public input, and based on a sectoral or segmented view of the world, one that may not accord with how real-world AFSs function. For example, in the UK, there are 16 departments or agencies involved in the regulation of the AFS and its relationship to agriculture, trade, health, food safety, and the wider economy. Therefore, insufficient efforts are typically made for policy development within a ministerial silo to be assessed holistically.

Innovations in this space include attempting to harvest "the wisdom of the crowd" through citizen assemblies, which take their form by bringing groups of citizens together to discuss issues to aid the navigation of complex policy space, and for more deliberative attempts to develop cross-government policy solutions. One nice example is the UK's National Food Strategy, which aims for a holistic approach to AFSs, seeking system-positive outcomes, rather than outcomes that simply improve farm profitability, reduce food prices, tackle food waste, or attempt to carry out some other narrow goal.

A key area for perhaps riskier and more transformative policy framing would be the use of alternative metrics of national good, beyond national income. The level and/or rate of growth of GDP or national income—typically in inflation-adjusted, per capita terms—are too often regarded as the principal socioeconomic performance index. The deficiencies of GDP and income measures are very well known: they fail to internalize damages to nature or climate; they do not value essential nonmonetized activities (e.g., caregiving) but do value monetized destructive activities (e.g., weapons manufacture and sales); and beyond a low-level production and consumption, measures correlate weakly, if at all, with happiness and life satisfaction measures (Stiglitz et al. 2010). At a societal level, which depends on living on a finite planet, incentivizing consumption growth also drives unsustainability, which increasingly—such as through climate anxiety—is undermining well-being. Innovative ways to measure societal well-being in a more nuanced way than GDP have been developed (e.g., the OECD's Better Life Index [OECD 2018]), and a few countries have begun to use these measures in place of GDP and national income. Improved measures shift discourse in a system-positive direction. These would incentivize a very different innovation environment, as the contribution of diets to planetary health and well-being would require that they not be considered "externalities" to the AFS.

Citizen assemblies might also inform the navigation of complex systems, where trade-offs are rife, and lead to greater clarity as to what positive systemic outcomes citizens really want, rather than relying on simple, unnuanced, and inaccurate proxies like "consumers want cheap food." Lack of clarity presently facilitates special-interest-group capture of vast public agri-food sector subsidies, mostly in forms that distort markets and in a manner that aggravates pre-existing income and wealth inequality because the wealthiest farmers—indeed, often just landowners who do not farm—receive the biggest payments (OECD 2020). **It is hard to believe that present subsidy programs reflect citizen desires as opposed to raw interest group pressures**. With greater clarity around desired systemic outcomes, public money can be more effectively targeted at them. As highlighted by Springmann (2021), changing the subsidy regime can have significant impacts on food availability, price and diets, and dietary illness (e.g. refocusing subsidies from calorie-dense starchy grains to fruit and vegetables).

Sustainable Animal and Plant Production Systems

Innovations in this sphere range across scales from microbiome-related advances that entail interactions with plants and animals to genetic technologies applied to microbes, animals, and plants to agronomic and other systems-management innovations. Many of these innovations are most effective when combined, potentially leading to synergies enabling novel syndromes of production (Vandermeer 1997; Finckh 2008; Li et al. 2020a). As outlined above, innovations in genetics, breeding, and agronomy have contributed to huge improvements in productivity over recent decades, with a range of downsides that are increasingly recognized as unsustainable.

Crop innovations: The power of conventional breeding continues to benefit from myriad innovations that are collectively termed molecular breeding, which encompasses a range of technologies from identifying natural gene variants (alleles) of interest and selecting for them in breeding programs, to moving genes within or across species barriers through genetic engineering, as well as gene editing to alter alleles within a genome (Jaganathan et al. 2018). Crop improvement strategies with implications for food and nutrition include those that increase yields, those that stabilize yields and reduce losses, and those that change the nutrient content of the crops. Past yield increases have resulted from hybrid technologies, in which the superiority of the progeny of crosses between certain inbred lines has allowed for astonishing yield improvement over recent decades, most notably in maize (Reeves and Cassaday 2002). Breeding new crop varieties with high nutrient-use efficiency is an effective means to reduce fertilizer use without sacrificing crop yields (Shen, Li, et al. 2013; Jiao et al. 2016). Most often, this has enabled yield increases with stagnant fertilizer use patterns, as has been true for maize/corn in the US. The idea of improving yields through improved photosynthetic efficiency (e.g., C4 rice) is a long-shot bet in which substantial research resources are being invested (Ermakova et al. 2020). Another high-risk idea with potentially high returns is that of putting genes into cereals to enable the crops to fix nitrogen (Box 2).

For stabilization of yields, there are many natural alleles and transgenic strategies aimed at improving abiotic and biotic stress tolerance. Ensuring that plants and animals can withstand challenges from biological aggressors is an inherently dynamic challenge because weeds, insects, and pathogens all evolve to overcome the obstacles that they encounter. For example, crop germplasm (including the wild relatives of cultivated crops) often carries a wealth of resistance alleles that can protect crops from diseases and insect pests when the alleles are transferred to cultivated varieties. Unfortunately, genetic resistance often breaks down rapidly when pathogens and insects evolve to evade recognition or otherwise overcome defenses (Pilet-Nayel et al. 2017); such boom-and-bust cycles also occur with pesticides. Certain forms of resistance are harder to overcome than others, and some forms of resistance may come with tradeoffs related to yield or vulnerability to other biotic stresses. For example, breeding plants that resist mycotoxins can often lead to low-yielding varieties, and there is an urgent need to develop varieties that are resistant to toxins based on mechanisms that do not reduce yields. Elegant strategies have been devised for engineering resistance, either with gene transfer or editing, but the implementation of these strategies has generally been complicated by public concerns about—and resulting heavy regulation of—genetic engineering (see Box 2). While diverse pest management options abound, from breeding to biological control to landscape management, pesticide use remains a global and generally toxic default paradigm.

Similar innovations enable crops to more efficiently use nutrients like nitrogen and to cope with soil stresses such as water deficit, salts, and other abiotic challenges. Improving tolerance to transient flooding through the introduction of a gene sourced from a rice variety adapted to deep waters has been an impressive success story in rice (Bailey-Serres et al. 2010; Oladosu et al. 2020). While single genes with such strong effects on stress tolerance are relatively rare, most traits can be modified by conventional breeding strategies that change the allele frequencies of multiple genes with small effects on the trait. A diversity of approaches are being used by rice researchers worldwide to identify the genetic variation among domesticated rice species and their wild relatives that can be exploited to breed a new generation of "green super rice" varieties (Wing et al. 2018).

In addition to increasing and/or stabilizing food production, innovation in genetics and breeding has contributed to improving culinary and/or nutritional quality (or to reducing quality, when these aims are not considered). Biofortification aims to improve the nutritional quality of foods by improving plants' vitamin, mineral, and/or fatty acid profiles. This can be achieved based on genes that influence the levels of nutrients in food, such as increased levels of pro-vitamin A with transgenic golden rice or with conventionally bred orange-fleshed sweet potato. Improved nutrient content (or increased bioavailability and/or decreased anti-nutrient content) can also be achieved by selection for these traits in conventional screening and/or breeding programs.

Box 2: Transgenic and gene editing technologies

Following their commercial introduction in 1996, transgenic crops ("GMOs") are now grown on more than 190 million hectares (ISAAA 2019). The vast majority of transgenic plants currently grown contain only a few transgenes that contribute to pest management through herbicide tolerance and/or insect resistance genes. Plants expressing toxin genes from the bacterium *Bacillus thuringiensis* have been widely deployed, resulting in both crop-yield improvements and reductions in pesticide applications (Brookes and Barfoot 2018; Pixley et al. 2019). A variety of other strategies for engineered pest resistance have also been developed (NASEM 2016; Talakayala et al. 2020). A wide variety of technically effective transgenic methods exist for managing virus diseases, for example, some of which have attained some commercial success (Pixley et al. 2019). Papaya ringspot virus (PRSV) resistance was commercialized in Hawaii in 1998 (Gonsalves 1998). Transgenic papaya for PRSV resistance was approved for commercial cultivation in southern China in 2006 (Li et al. 2007) and has been planted there on a large scale. There are many other strategies for producing disease and insect resistance through gene transfer, which can be combined in transgene cassettes carrying multiple resistance genes (van Esse et al. 2020). For example, trials are being conducted in Uganda to assess potato lines carrying multiple resistance genes against late blight (Ghislain et al. 2019).

The traits in most widely cultivated transgenics target production-related priorities, mainly pest and weed management. Additional transgenic approaches have shown encouraging results for enhancing other production-related traits, such as drought tolerance (e.g., NASEM 2016; Gonzáles et al. 2020). Traits that more directly benefit consumers include those that are nutrition-related, such as vitamin, mineral, or fatty acid contents. The production in plants of omega-3 long-chain polyunsaturated fatty acids, normally sourced from fish oils, could reduce the pressure on oceans to supply this important nutrient that is often limited in diets. Products from the transgenic crops, carrying genes from marine microbes, can be consumed by livestock, fish, or humans. A number of other genetic innovations are currently in the pipeline with a focus on nutritional enhancement of crops and livestock (NASEM 2016; Napier and Sayanova 2020).

Extending the range of plants that can capture ("fix") atmospheric nitrogen (N) could benefit the environment by reducing the unsustainable production and use of synthetic N fertilizers (Charpentier and Oldroyd 2010; Galloway et al. 2013; Van Grinsven et al. 2013; Ladha et al. 2020). Natural substitutes, such as leguminous crops (beans, peas, and similar), can utilize N from the environment through symbiosis with root-associated bacteria that can fix N. Other types of plants, such as the major cereals, cannot form such productive relationships. Increased use of N-fixing crops would reduce energy and GHG emissions arising from fertilizer manufacturing. It could also reduce nutrient (especially N) losses into the air and water that contribute to both pollution and climate change.

Engineering crops, especially staple cereals, to fix N is a long-standing aim toward which plant breeders have made significant advances in recent years, although no new variety is anywhere near ready for widespread release. Several strategies are being undertaken to enable this, such as transferring the genes that control the development of root nodule symbiosis from legumes to cereals; creating nodule-independent N-fixing cereals promoting their association with endophytes that fix N; gene editing of associative N-fixing bacteria; and directly introducing nitrogenase into the plant (Mus et al. 2016; Vicente and Dean 2017; Rosenblueth et al. 2018; Van Deynze et al. 2018; Bloch et al. 2020). The tradeoff is that N fixation is metabolically expensive for legumes because they "feed" their symbionts carbon in return for the N fixed, so N-fixation would likely constrain yield potential in cereal crops. If the resulting yield reductions compel expansion of the agricultural frontier, resulting in the conversion of forest to croplands, the net environmental impact of N-fixing cereals could be adverse. It therefore remains to be seen whether, and when, cereal yields and associated environmental impacts could be enhanced by incorporating N-fixing capability in the absence of applied N.

Despite the commercial success of a few categories of transgenic plants, the approach remains controversial (Chvátalová 2019). As a consequence of this and the expense associated with clearing regulatory hurdles (US$7–35M out of a typical product development cost of US$136M; Phillips McDougall 2011), the transgenic crops in commercial cultivation use only a very small proportion of the genetic variation that could be accessed through this approach.

The more recent emergence of genome editing technologies (i.e., CRISPR/Cas), has made it possible to precisely alter gene sequences native to an organism. This is being widely applied to plant and animal species, as a powerful tool for genetics and breeding that may obviate some transgenic approaches (van Eck 2020; Mao et al. 2019). This technology can contribute to crop diversification by allowing rapid improvement of key agronomic traits in hitherto neglected crops, as recently shown by leveraging insights from tomatoes to improve plant architecture and fruit size in groundcherry (Lemmon et al. 2018). The regulatory environment for the utilization of this technology in agriculture—including the ethics of gene regulation—will largely determine the extent to which this technology contributes to crop diversification, protection, adaptation to climate-related stresses, and nutritional quality (Zaidi et al. 2019; Smyth 2020).

There seems no impending shortage of genetic engineering applications to effectively tackle disease resistance, abiotic stress tolerance, and desirable consumer attributes; the constraints to development, diffusion and equitable impacts largely stem from social and institutional forces (Pixley et al. 2019).

To date, the focus of genetics, breeding, and seed systems has been on the deployment of high-yielding starchy staples, with some shift of focus to more diverse foods in recent years. Even within the starchy staples, there has been an extreme emphasis on a few crops, which can make for vulnerability to climate events, pests, diseases, and other stressors. **Further investment in research beyond the major cereals could contribute to diversified cropping systems** that are better adapted to a range of environments (DeFries 2018; Mason-D'Croz et al. 2019). Focusing more research and innovation on a much larger range of plant and animal species could support the strategic diversification of food production and consumption, and help address the reduction in agrobiodiversity that has come with the focus on a narrow range of species. For example, more research on trees that produce fruits and nuts, as well as diverse vegetables and other more nutritious and sustainable food sources, could contribute to more resilient production systems and better diets.

Creating adapted and stress-tolerant germplasm is one set of challenges; ensuring that farmers have the germplasm they need is another. Seed value chains remain severly underdeveloped in many low-income countries (Ariga et al. 2019; Barriga and Fiala 2020). Facilitating the emergence of viable, reliable seed value chains is an essential first step in promoting adaptive genetic improvement research and farmer uptake of those improvements. Innovation in varietal evaluation and seed systems includes old and new strategies for working with large numbers of farmers to test varieties in diverse contexts (e.g., Bänziger and Cooper 2001; Van Etten et al. 2019). Plant breeding is, in any case, an important but relatively small component of the socio-technical strategies needed to build the climate resilience and sustainability of food systems. For farmers to implement more sustainable production practices, the innovations must be developed; farmers must be aware of them and have the knowledge, skills, and technologies needed to implement them; and producers must actually change their behavior. Experiences in the commercial, public health, and agriculture sectors illustrate that interpersonal contact can be essential in driving large-scale behavior change (Gawande 2013). Large-scale agronomic studies are being conducted in

China based on the "Science and Technology Backyard" (STB) system that links smallholder farmers with extension and research through a village-level innovation platform (Box 3), and elsewhere using farmer research networks and related approaches (Nelson et al. 2019; Van Etten et al. 2019). These methods show tremendous promise for drawing together the wisdom of (small farmer) crowds with the knowledge of cutting-edge scientific researchers to accelerate discovery, adaptation, and diffusion.

Box 3: Science and technology backyards—linking farmers, extension, agribusiness, and science at scale[8]

In China, tens of millions of small-scale farmers have implemented resource-conserving and yield-enhancing farming techniques through the STB initiative. The STB approach also gives researchers large datasets in near-real time to establish what works for whom through the participatory research and extension built into the approach. STB began with the observation that established training methods were not leading to substantial smallholder adoption of innovative cultivation methods; few farmers changed their practices even if they knew of the technologies. This recognition inspired China Agricultural University (CAU) researchers to strengthen their engagement with farmers. Recognizing the key role of trust, and understanding the importance of two-way information flow to support agroecological intensification, CAU scientists began a participatory research and training effort in Quzhou in 2009.

The researchers moved their research programs from the experimental station to the village so they could work and communicate directly with farmers, to share in their successes and failures. They rented a backyard in the village, and lived, worked, and studied in the yard. Professors and postgraduate students conducted intensive farmer participatory trainings. Gradually, farmers were attracted to the backyards, which became science and technology dissemination focal points in local communities. Trained farmers adopted high-yield and high-efficiency technologies (e.g., formulated fertilizer, sowing technology, and efficient water and fertilizer use techniques) at much higher rates than did untrained farmers (Shen, Cui, et al. 2013; Jiao et al. 2019). The effort then scaled up dramatically.

[8] We thank Xiaoqiang Jiao and Fusuo Zhang for contributing to the content of this box.

STB is now a multi-actor innovation platform located in rural areas that links the scientific community with smallholders, local government, and private enterprises to facilitate information exchange and technological innovation for achieving sustainable intensification of agriculture. The platform consists of farmer field schools, participatory on-farm research, new technology demonstrations, and farmer interest groups or clubs. Farmers get rapid and context-relevant responses to their challenges. Companies contribute their technologies and funding, quickly learning what works and what doesn't. Local governments provide supportive policies and extension services, earning constituent support.

In 2020, 127 STBs operate in 23 provinces and regions with the participation of 29 scientific research institutes and over 100 agricultural extension stations in China. The STB system covers 45 major crops and has allowed significant scientific insights to be made, while facilitating transformative change that both improves yield and decreases the environmental footprint of agriculture (W.F. Zhang et al. 2016; Jiao et al. 2019). In 2019, FAO partnered with CAU and African countries to promote STB for enhancing transformation of African agriculture, starting with 34 students from eight African countries training at CAU before returning to their home countries to implement STBs (Jiao et al. 2020).

Livestock innovations: The livestock sector is often blamed for contributions to communicable and non-communicable disease burdens and to greenhouse gas production. But while reduced meat consumption is recommended in industrialized food systems, greater meat consumption would be beneficial to health outcomes in many low- and lower-middle income countries (FAO 2020). Tens of millions of resource-limited households derive their livelihoods from livestock and improving productivity in the sector can contribute to improving nutrition and pro-poor development in general (ILRI, 2019). Many actions can boost productivity including improved grazing, better disease management, and closer integration with other on-farm enterprises such as crop production. Two areas of innovation are highlighted below: improved livestock breeding and feeds.

Sophisticated livestock breeding methods have been applied to improve livestock productivity. Advanced genetic and genomic selection methods have the potential to contribute to heat tolerance and to methane mitigation (Pryce and Haile-Mariam 2020; more on the latter issue below). Livestock breeding efforts that focus on other production traits tend to reduce heat tolerance, which is problematic as temperatures rise with climate change. This trend requires attention to breeding for heat tolerance. An example is the "slick hair" trait, which increases thermotolerance and productivity in Holstein cows (Ortiz-Colón et al. 2018). While prospects exist for accelerating traditional breeding processes for desired animal traits (Strandén et al. 2019; Barbato et al. 2020), an integrated approach will require both technical and social adaptations (Menchaca et al. 2020). Many indigenous livestock breeds and populations remain uncharacterized, particularly in Africa, and much is unknown about their cross-breeding potential. Increasing the attention focused on a wider diversity of locally adapted species, including small stock such as guinea pigs, sheep and goats, may increase production in niches important to the food security of vulnerable populations.

Innovations in feed value chains can address a range of AFS dysfunctions. Examples include feed-based strategies for reducing methanogenesis in ruminant digestive systems to reduce greenhouse gas production in the livestock sector; reducing the depletion of fisheries stemming from the use of fish-based fish food; and improving the levels of omega-3 in animal and human diets. Algal-derived feed supplements can be used to substantially reduce enteric methanogenesis in ruminants (McCauley et al. 2020). Furthermore, synergies have been observed between the effects of algal biomass on methane production and livestock productivity.

Another innovation is the use of insects as feed. Insects are often rich in protein and some vitamins and minerals. In the EU, black soldier fly (BSF), yellow mealworm and the common housefly have already been identified for potential use in feed products (Henchion et al. 2017). Use of some insect-derived protein may reduce GHG emissions, though strong evidence on this impact remains scant (Parodi et al. 2018). Insect-based feeds are currently advanced mainly for their nutritional, environmental, technological and socio-economic impacts.

Consumption of fish and shellfish is recommended for personal and planetary health (Willet et al. 2019). A variety of innovations are improving the prospects for sustainable production of these foods. As world fisheries decline with increased anthropogenic and climate stress on the world's oceans, aquaculture has become an increasingly important source of fish and shellfish, especially in the Global South. Production from low- and middle-income countries in South Asia, Southeast Asia, and Latin America is increasingly responsible for the growth of global aquaculture and shows considerable future promise (De Silva 2012; Gentry et al. 2017). Well-designed aquaculture systems can deliver nutrient-rich foods with low environmental impact (Shepon et al. 2020).

Much of the ocean's fish catch is used to feed farmed salmonids (salmon and trout) and shrimp. Shifting away from the inclusion of fish meal in aquafeed could enable aquaculture farms to produce high-value products like salmon and shrimp without depleting the ocean's fisheries or expanding current, less sustainable feed-cropping systems such as soy and canola (Fry et al. 2016). Options to reduce the environmental footprint of fish feed include insects, such as BSF and algae, both of which can be grown using side products (i.e., potential wastes; see Box 4). Likewise, single cell proteins (SCPs) produced via fermentation are also ideal fish meal substitutes, and some use methane as feedstock, making them even more sustainable. Another option is camelina, an oilseed crop that can be used as an animal feed to enhance omega-3 levels (Berti et al. 2016). In addition, it is well adapted to genetic manipulation and so can also be used to produce very high-value lipids (Yuan and Li 2020).

The genetic diversity of farmed fish is currently low, and pests and diseases may be poorly controlled in ways that are harmful to the environment and human health (e.g., Cabello et al. 2013). The diversification of aquatic species used in aquaculture could reduce pest and disease pressure and provide a wider range of options for cultivation in different environments. New aquaculture production models are emerging to tackle environmental issues such as eutrophication and mangrove loss, including land-based recirculating aquaculture systems (RAS), inland coastal flow-through systems for salmon and indoor farms for shrimp, and ocean-based closed containment systems. RAS is an especially promising technology, offering the potential to grow seafood entirely indoors with minimal environmental impacts. In these systems water is continuously reused, and fish waste, uneaten feed, nitrates, and microorganisms are filtered out. Current species approaching commercialization potential include salmon, trout, tilapia, kingfish, barramundi, and shrimp.

Agroecological innovations: The varied challenges created by modern agriculture can be addressed, at least to some extent, by a shift from reliance on hydrocarbons-based inputs to the application of approaches that are based on agroecological principles such as efficiency, synergy, and circular economy (Barrios et al. 2020). Similar concerns and approaches are described in literatures associated with the terms "regenerative agriculture" and "agroecology." The field of agroecology (AE) entails "the study of the interactions between plants, animals, humans and the environment within agricultural systems" (Dalgaard et al. 2003). The term AE is also used to refer to the science, practice and movement related to the ecological and social processes that underlie and influence farming and AFSs (Wezel and Soldat 2009). Holistic approaches to AE consider both technical and social levels through interconnected innovations that can work together to transform food systems towards greater sustainability (HLPE 2019). The concept of agroecological transition has been highlighted in a number of recent reports and case studies (IPES-Food 2018; NatureScot 2020).

Gliessman (2016) outlined a series of stages that can support a transition to ecologically based agriculture. The first stage is raising efficiency. Increasing input use efficiency is a major focus of "sustainable intensification" (e.g., Godfray et al. 2015). Important improvements in nutrient-use efficiency are being achieved by precise placement of designed nutrients in the root zone, with the quantities, composition, and timing of application guided by models (Shen, Li, et al. 2013; Wang and Shen 2019). This approach can support increased yields via root-soil-microbial interactions (Wang and Shen 2019; Wang et al. 2020). Strategic nutrient application has also been coupled with intercropping to improve yields and nutrient use efficiency (Li et al. 2007). A range of integrated soil fertility management (ISFM) practices similarly couple use of external inputs as essential complements to more effective use of organic materials from within the system (Place et al. 2003; Vanlauwe et al. 2010, 2015).

The second level of agroecological transition involves substituting natural processes for excessive use of chemical inputs. This includes the integration of legumes into cereal-based systems to bring in nitrogen, or crop-livestock integration as another alternative source of crop nutrients. While modern agriculture has relied on toxic chemicals to manage pests (i.e., insects, weeds, and pathogens), ecologically friendly management options have been, and are being, devised especially to combat emergent pesticide resistance as insects and pathogens evolve in response to chemical controls. "Biological control" can involve the use of native natural enemies, encouraged through landscape management, as well as introduced predators of pest species and microbial antagonists of pests. Spectacular outcomes have been achieved towards managing the cassava mealy bug (Herren and Neuenschwander 1991) and the pearl millet head miner (Ba et al. 2014), for example, through the introduction of parasitoid insects that prey on the pests, and there are many new possibilities for biological control (van Lenteren et al. 2018). Box 4 focuses on regulatory approaches that could break the current pesticides lock-in.

The third level of agroecological transition entails redesigning production systems to avoid problems and drawing upon new agroecological principles and processes (Krebs and Bach 2018; Pretty et al. 2018; Barrios et al. 2020; Wezel

INCREASING DIVERSITY CAN PROVIDE A RANGE OF BENEFITS THAT COLLECTIVELY IMPROVE SYSTEM RESILIENCE.

et al. 2020). This may entail new crops, as well as the integration of crops, trees, and livestock, and nutrient flows between rural and urban areas. **Increasing diversity can provide a range of benefits that collectively improve system resilience**. For example, a long-term study of diversification via crop rotation showed that maize yields were higher with more diverse rotations, even under drought conditions (Bowles et al. 2020). Although crop diversification is hardly a novel idea, shifting from monocultures and other low-diversity systems towards greater agrobiodiversity may involve innovation in breeding, agronomy (potentially including engineering), and markets (IPES-Food 2016). In a variety of contexts, redesign of integrated crop-livestock systems can offer environmental and economic benefits. For example, Bonaudo et al. (2014) cite examples of successful application of agroecological principles towards improving system performance through crop-livestock integration in Brazil (reducing deforestation in the Amazon) and in France.

System redesign will entail diversifying production systems in time and space, considering the integration of crops, livestock, and trees—as well as external inputs—on farms and across landscapes. Diversity at these scales has implications for nutrient cycling, natural pest regulation, risk management, and, in some contexts, the diversity of consumption. Modern agriculture has too often reduced diversity; reversing this trend will require new approaches to landscape management, taking into account the interests of multiple stakeholders and the ecosystems services they require (Moraine et al. 2014; Martin et al. 2016). At the same time, we must guard against overcelebration of diversity as an end unto itself lest we risk locking in a low productivity status quo among smallholder producers who need external inputs, and perhaps greater partial specialization in order to escape poverty. The point is the need to tap the best insights of both the AE and sustainable intensification approaches and to customize solutions to specific contexts rather than paint with too broad a brush.

Plant and animal breeding can be regarded as combinatorial genetics; breeders use recombination and selection to put together not only the best alleles but also the sets of alleles that harmonize best with each other. Similarly, combinatorial agronomy has the potential to more fully exploit the interactions of genotypes with environments, as well as the interactions of multiple crop varieties and species in diversified systems. Plant varieties and species can synergize based on complementarity of resource use, as well as less obvious biochemical interactions (D.S. Zhang et al. 2016; Wen et al. 2019; Zhang et al. 2020). In addition, the performance of plants and animals can be influenced by the microbes associated with them. For example, certain root-associated microbes can enhance nitrogen fixation in legumes, and others can benefit wider ranges of plant taxa.

Growth-promoting Rhizobacteria, for example, can greatly enhance the performance of potatoes grown under biotic and abiotic stresses (Grossi et al. 2020). The effective design and implementation of biodiverse landscapes is a combinatorial challenge that, unlike plant breeding, cannot easily be conducted by the private sector alone. Large-scale public engagement in innovative-farming system design can be facilitated by digital technologies and collaborations such as the Science and Technology Backyard (W.F. Zhang et al. 2016; Box 3) and the farmer research network approach (Nelson et al. 2019).

Diversified crops and systems, together with markets that support people's access to diverse foods, provide a wide range of options for improving dietary nutrient intake. There is a complex relationship between the diversity of production and the diversity of diets, but there seems to be a strong positive relationship between agricultural biodiversity and dietary diversity in smallholder systems (Sibhatu et al. 2015; Jones 2017a, b; Sibhatu and Qaim 2018; Tobin et al. 2019). Diversity can offer a variety of important ecosystem services, from reducing epidemic potential (King and Lively 2012) to enabling different species to tap soil, water, and light resources in complementary ways that improve yields (Li et al. 2020b; IPBES 2019). Plants can also be biofortified based on improving fertilizer use, either in the field or in controlled-environmental contexts; this does not require genetic modification (Pannico et al. 2019). Levels 4 and 5 of Gliessman's framework for agroecological transitions entails reestablishing connections among those who produce and consume food, and building a new global food system based on greater equity and justice (Gliessman 2016). Much of this report focuses on the mechanisms that could deliver on the HERS objectives that are shared by those in the agroecology movement and by others who may have different foci and couch their arguments in different language.

Box 4: Regulatory nudges towards integrated pest management

A substantial proportion of crops are lost to pests, which are broadly defined as including the weeds, microbes, and insects that reduce yields (Oerke 2006). Despite many well-known downsides and alternatives (Sánchez-Bayo and Wyckhuys 2019), pesticides remain the global standard approach to managing pests. The effectiveness of the pesticide solution is showing signs of wear. Over the past several decades there has been rapid increase in the evolution of biological resistance to crop protection compounds (Gould et al. 2018), and concern is growing over the environmental and health impacts of pesticides. There are a wide variety of integrated pest management (IPM) alternatives, though some IPM methods may be more complex to operationalize than spraying pesticides. Regulatory pressure is needed to enable a general shift from synthetic pesticides to agroecologically-based IPM approaches.

Regulatory frameworks for the crop protection industry currently vary greatly, with the strongest regulation in Europe and the weakest in many LMICs. Most current regulations in the US focus only on the active ingredient, while a growing body of research has shown that other components of product formulation can be as toxic as active ingredients (Benachour and Séralini 2009). Because agricultural intensification in LMICs has the potential to increase the use of pesticides, regulatory environments need to be strengthened in these areas to ensure the safety of workers, consumers, and the environment. Farmers in Africa are increasingly dependent on pesticide use, with associated human health costs (e.g., Sheahan et al. 2017).

Regulations should require that products be able to be used safely. In low-resource contexts, farmers may lack access to personal protective equipment (PPE) necessary to manage chemicals safely, or existing PPE may be inappropriate for use in local environments—such as heat-trapping slickers used in hot, tropical environments where temperature increases are known to have physiological effects on workers (Masuda et al. 2020). Also, many potential users may not be literate in the languages in which safety guidance is provided. Some crop protection companies have committed to the improvement of training for farmers, but requirements should be introduced such that registration of a product in a market segment is not allowed unless it has been clearly demonstrated that most farmers can use it safely.

Many agroecological principles and practices can be used to manage pests without the use of synthetic pesticides or in a manner that can at least sharply reduce pesticide use. The use of host plant resistance is already widespread, though seed companies that benefit from pesticide sales tend to focus on improving yield potential rather than resistance in their breeding programs. The use of biodiversity (greater diversity of crop species and varieties, strategically deployed in time and space) can reduce pest pressure and epidemic potential (McDonald 2014), while potentially contributing to system resilience and dietary diversity. Botanical pesticides (e.g., chemicals derived from plants—often local weedy species) have proven to be useful in pest management even in very low-resource environments, often reducing pest populations without harming the pests' natural enemies (Stevenson et al. 2020; Sola et al. 2014). Biological control agents, including microbial pesticides, can be very specific and effective (van Lenteren et al. 2018; Lednev et al. 2020).

The crop protection industry's current profit model is based on the volume of product sold. A shift in the business model to provide a service—pest management—rather than a product could push the industry to develop new mechanisms for monitoring and managing pests. In such a model, chemical inputs would be a cost to the service, rather than the primary source of revenue. Promising IPM approaches would likely be amplified because of their potential lower cost per acre. There would also be incentives to target that limited chemical use to specific locations in a field and at specific times to minimize the development of resistance. The industry could license digital tools that identify, track, and provide targeted recommendations for sub-field pest management approaches.

Soil health innovations: Soil health poses a fundamental challenge to agriculture in all AFSs. The application of mineral fertilizers can temporarily obviate productivity constraints posed by specific nutrient deficiencies (especially N) and can, but does not always, support the maintenance of soil organic matter (SOM), which is fundamental to soil health. The organic component of soil is especially critical to its structure, ability to cycle nutrients, resistance to erosion, regulation of hydrological processes, facilitating recharge, and water holding capacity. A rule of thumb holds that for every 1 percent organic matter in soil per hectare, 100,000 liters of water can be held by the soil. SOM depletion is an especially severe threat to much of Africa and parts of South America, where soils are ancient and weathered and SOM has been depleted to the point that it cannot support crop growth. Strategies to boost SOM must be adapted to local soil conditions and management options, sometimes requiring increased external inputs of inorganic nutrient amendments and in other cases reduced application rates (Amelung et al. 2020).

Challenges related to soil, water, and climate are interrelated, so their solutions need to be considered and approached in an integrated way. SOM is one of the earth's main carbon pools. It can either sequester or release carbon, and thus has a key role in the global dynamics of GHGs. The French government has announced the aspirational "4p1000" initiative, on the premise that climate change could be halted if carbon were returned to soils at an annual rate of 0.4 percent. Climate change is working against us. The increasing temperatures associated with global warming make organic matter less stable, and violent rain events contribute to soil erosion and loss of organic matter. Land management approaches that build and maintain soil carbon can both reduce GHG emissions and sustainably improve food and feed production. These approaches include landscape management to reduce erosion, including agroforestry; and the use of cover crops, especially leguminous species that fix nitrogen and access poorly soluble forms of phosphorus by carboxylate exudation from roots (Lyu et al. 2016; Griscom et al. 2017; Wen et al. 2019). A challenge, however, is that more marginal lands and soils—where a disproportionate share of the rural poor reside—cannot sequester much carbon per unit area. Investments to build soil carbon can thereby inadvertently exacerbate economic inequality without companion interventions to help those in marginal areas.

An especially promising option is increased use of carbon and other nutrients recovered from organic waste, including food waste, industrial waste (e.g., coffee cherry, sugarcane bagasse, sawdust, animal bones), as well as human and animal waste (urine and feces). The volumes involved, and the negative health and societal effects of these waste streams, are enormous, as are the potential benefits of recycling and reuse (Berendes et al. 2018; Mihelcic et al. 2011). Many sources of organic matter currently lead to GHG emissions and air and water pollution (e.g., nutrient loading of aquatic environments causes toxic algal blooms). Human, animal, and other organic wastes have historically been used as fertilizers, but these practices have eroded for a variety of good reasons for which there are now technical solutions.

The recovery of nutrients and organic matter from waste streams into agriculture could provide a wide array of benefits—including improved sanitation, reduced pollution, reduced GHG burden from agriculture, improved climate resiliency, improved crop production, and improved health of soils and people—contributing to achievement of most SDGs (Orner and Mihelcic 2018). Sewage sludge has fertilizer value but also the potential to contaminate soils and foods because of industrial wastes that can enter sewage systems. A large fraction of humanity is not served by sewers in any case; 2.3 billion people lack even basic sanitation, and the excreta of 4.5 billion goes into the environment untreated (WHO and UNICEF 2017). Container-based sanitation can circumvent these problems (Russel et al. 2019). Technical developments can facilitate resource recovery into agriculture, with a diversity of options that can be adapted to different contexts (Harder et al. 2019). No single step in waste-to-value chains is prohibitively challenging, but large-scale implementation would require considerable political will and social adjustment to overcome a range of barriers to implementation.

A key barrier is the distance between the urban locations where most waste originates and the peri-urban and rural loci of most agricultural production, which often makes transport of low value-to-weight waste products prohibitively expensive. This will often require low-cost processing to increase the value density of recovered waste by products. Such processing techniques are themselves important areas in need of innovation.

A key challenge associated is that improving soil health is a slow-moving process, with many soil features being non-obvious and longer-term. Innovations in soil health include the development of toolkits that enable people (farmers, ranchers, and other land managers) to discern what is happening to soils more quickly, cheaply, and reliably. Conventional soil testing is done in laboratories, using methods that require considerable time, expense, and expertise. New methods allow more farmer-friendly assessments, as well as remote sensing of soil features (Magonziwa et al. 2020). Several teams are developing lower cost, higher spatial resolution methods to assess soil chemistry (organic constituents, inorganic macro- and micronutrients, pH, etc.), biology (microbes and macrofauna) and physics (structure, including aspects that influence water infiltration and absorption). The use of spectral methods, including those that utilize low-cost, hand-held spectrometers, as well drone-based and satellite-based ones, is rapidly bringing down the cost and speed of access to soil-related data (Angelopoulou et al. 2020). Spectral methods must be complemented by digital soil mapping that leverages site-based measures, satellite-based covariates, and artificial intelligence methods to predict at scale soil properties (e.g., micronutrient availability) that are not especially amenable to detection by spectral methods.

The capacity to generate reliable, affordable, farm- or field-specific soil indicators and fertilizer recommendations remains limited, however. Errors inevitably arise in soil sampling and chemical analysis procedures within and among laboratories, and in algorithms predict soil conditions based on a few imperfectly observable indicators. The marketing of technology-based advances in soil information services currently overreaches their capacity to deliver (Schut and Giller 2020), much as has been

DIVERSIFICATION OF PRODUCTION CAN CONTRIBUTE TO HEALTHY DIETS BY PROMOTING DIETARY DIVERSITY, ESPECIALLY AMONG SEMI-SUBSISTENCE SMALLHOLDER FARMERS.

broadly the case for index insurance products intended to provide farmers with low-cost, context-specific risk management products (Jensen and Barrett 2016).

Sustainable production-system innovations connect to several of the food-system design objectives outlined above. Perhaps agroecological intensification's chief benefits relate to environmental sustainability: curbing the expansion of farmed land at the expense of nature, as well as reducing pollution associated with the indiscriminate application of fertilizers and pesticides. **Diversification of production can contribute to healthy diets by promoting dietary diversity**, especially among semi-subsistence smallholder farmers (Sibhatu et al. 2015; Sibhatu and Qaim 2018). Improving soil health can improve resiliency to drought spells, thereby improving the productivity on marginal (and other) lands (Lal 2016).

By reducing plant stress, healthy soils can also reduce a widespread food safety issue: the problem of mycotoxin contamination. Mycotoxins are toxic metabolites produced by a range of micro-fungi that colonize foodstuffs (most notably, maize, groundnuts, tree nuts, and spices) before and after harvest. The best known are aflatoxin and fumonisin, but hundreds of other mycotoxins contaminate the world's food system. Aflatoxin, the most potent naturally occurring compound known, causes liver cancer, growth stunting, and immunosuppression (Routledge et al. 2016); it contaminates a quarter of the world's foodstuffs at levels above regulatory limits, and up to 80 percent at detectable levels (Eskola et al. 2020). There are many interventions that can be used to minimize or manage mycotoxins along food value chains, but ensuring soil health is fundamental, as stressed plants are most vulnerable to fungal colonization. Crop genotype, as well as harvest and post-harvest conditions and processes, also influence mycotoxin accumulation and exposure. Boosting food safety commonly requires multiple complementary interventions.

Alternative, Land-Saving Nutrient Production Systems

At least four distinct, rapidly-advancing classes of innovation are already beginning to facilitate de-agrarianization. The costs of production in this space are dropping quickly as private investment pours into novel technologies that promise to reduce the land and ocean footprint of food production.

The first involves the emergence of nutrient-dense food and livestock feeds based on microalgae, insects (e.g., BSF larvae), etc., as substitutes for land-intensive cereals and oilseeds-based proteins and fish meal. The livestock sector accounts for 40 percent of the world's agricultural GDP and contributes to the livelihoods of 1.3 billion people (Herrero et al. 2013). Feeding animals also accounts for a large share of agriculture's environmental footprint. Assuming continued or growing demand for protein concentrate for livestock feed to meet rising demand for animal-source foods, alternative livestock feeds that utilize currently neglected resources could reduce the environmental footprint of meat and aquaculture production, while also reducing pollution from other sources and ensuring affordable and equitable access to these nutrient-rich foods.

Several multinationals have made strategic investments (often through collaborative ventures) in this field, with prominent examples such as Nestlé and Corbion,[9] as well as Unilever and Algenuity,[10] for microalgae food innovations; or Buhler and Protix[11] for insect-based food and feed. The unit-production costs of these novel alternatives are falling fast, and they should be able to compete commercially this decade with soymeal, maize, hay, fish meal, and other conventional feeds. Research shows that these feeds are scalable, yield animal-sourced foods of similar quality and safety as those based on conventional feeds, and potentially offer added health benefits (Caporgno and Mathys 2018; Smetana et al. 2019; Altmann et al. 2020; Cottrell et al. 2020).

Box 5: Microbial, insect, and algal biomass as circular feeds

Insects, themselves a miniature form of livestock (Barroso et al. 2017), have many advantages in feed value chains (van Huis 2013) for the rearing and maintenance of fish and shellfish, chickens, pigs, and pets. For example, black solider fly larvae (BSFs) can be fed organic wastes from industrial or

---

[9] See press releases at https://www.corbion.com/about-corbion/press-releases?newsId=2199459 and https://www.nestle.com/randd/news/allnews/partnership-corbion-microalgae-plant-based-products, or press coverage at https://www.foodnavigator.com/Article/2019/11/07/Nestle-and-Corbion-eye-microalgae-for-next-generation-plant-proteins#.

[10] See press coverage at https://www.foodnavigator.com/Article/2020/07/30/Unilever-and-Algenuity-discuss-the-potential-of-microalgae-Algenuity-s-technology-unlocks-a-wealth-of-food-applications#.

[11] See press coverage at https://www.feednavigator.com/Article/2017/06/27/A-new-Dutch-plant-will-be-the-first-in-Protix-and-Buhler-insect-tie-up.

municipal sources, such as food scraps and excreta from humans and animals (Gold et al. 2018), and used to produce the high-quality protein, fats, and other nutrients that are needed for livestock and humans (Patel 2019; Smetana et al. 2019). BSFs are tolerant of certain toxins, such as pesticides and mycotoxins, providing a disposal alternative for contaminated foodstuffs. Among other animal protein sources, BSF (either as a puree or meal) has a low carbon footprint and low potential for ozone depletion, acidification, and eutrophication impact (Smetana et al. 2019). The BSF market is projected to grow to more than US$2.57 billion by 2030 (Byrne 2020) and has already entered middle-income country markets such as Indonesia.[12]

BSF cultivation also has the potential to contribute to human waste management, thus providing an avenue towards achieving SDG 6, which concerns ensuring access to water and sanitation. Most human excreta and other organic wastes currently go untreated into waterways, with 92 percent of wastes being untreated in low-income countries and 80 percent untreated at a global level (Sato et al. 2013). The use of organic wastes for BSF production could improve water quality and safety while producing high-quality feed. A concern about BSF-based waste utilization relates to chemical and microbial safety. For example, the possibility of heavy metal contamination was demonstrated, as BSFs bioaccumulate heavy metals present in their diets. (BSFs have, thus, been considered for bioremediation [Bulak et al. 2018].)

Single cell proteins (SCPs) offer another high-potential source of nutrition for inclusion in feed for aquaculture and livestock. SCPs are protein meals based on microbial or algal biomass, and can be produced by yeast, bacteria, microalgae, and protists. These microorganisms generate proteins after consuming sustainable feedstocks including methane, wastewater, industrial and agricultural residues, methanol, syngas, and second-generation sugars. SCP manufacturers are scaling up operations globally, including commercial-scale plants in the developing world (Jones et al. 2020).[13]

[12] A BSF demonstration facility in Indonesia has completed final evaluation. See press coverage at https://www.eawag.ch/en/department/sandec/projects/mswm/forward-from-organic-waste-to-recycling-for-development/.

[13] See, for example, Calysta's entry into China https://www.undercurrentnews.com/2020/06/30/calysta-adisseo-aquafeed-joint-venture-to-build-first-plant-in-china/.

Microalgae are another valuable, well-rounded source of biomass, protein, oils, and minerals for aquaculture, livestock, and human consumption. Fish and fish oils are valued in human diets for their high omega-3 fatty acid contents, which are derived from the microalgae on which they feed. Sourcing these high-value oils directly from microalgae could reduce offtake pressure on marine fisheries, which are the main current source of fish meal and fish oil feeds.[14]

Lutein, a widely used carotenoid for food coloring as well as a dietary supplement, is sourced from microalgae. Lutein-rich spirulina microalgae (cyanobacteria *Arthrospira*) are used as a supplement for fish and human nutrition (Shah et al. 2018). Microalgae can be farmed in marine or closed-loop production systems to produce food and feed, while capturing nutrients that can otherwise damage aquatic resources. Contained production systems can be designed at varying scales, either with controlled lighting or in the dark with controlled carbohydrate inputs. However, due to limited technology readiness levels and economies of scale, both types of production systems are energy intensive and require substantial capital investment in many regions (Smetana et al. 2017).

The second class of innovations rely on tissue engineering methods that culture cells to grow animal tissue outside the body, without the environmental, animal welfare, or financial costs of raising and slaughtering live animals. These "clean" or "cellular" meats have attracted considerable private investment and media attention. The commercial threat these products pose to conventional livestock producers has already prompted legislative and regulatory battles in some OECD economies over product labeling (i.e., what constitutes "meat"). Although these products remain expensive, unit costs are dropping fast and are predicted to fall to the level of conventional ground beef by 2026 (Tubb and Seba 2019).

[14] In 2017, a joint venture between DSM Nutritional Products and Evonik Nutrition & Care was announced to invest around US$200 million in a new facility, delivering omega-3 fatty acid products for the fast-growing animal nutrition and aquaculture markets. See press coverage at https://www.nutritioninsight.com/news/DSM-Evonik-Collaborate-on-Marine-Algae-for-Animal-Nutrition-Aquaculture.html.

The third group of land-saving food innovations relies on controlled environment agriculture (CEA)—so-called "indoor" or "vertical" farming—much of it based on aero-, aqua-, or hydro-ponic methods. CEA is growing quickly to serve urban middle- and upper-class consumers in OECD and Asian countries. Its comparative advantage lies in year-round localized supply chains delivering consistent-quality, high-value, short-cycle horticultural products (Pinstrup-Andersen 2018; WWF 2020). Falling electricity costs and more reliable and affordable small-scale (e.g., rooftop) renewable energy generation increasingly obviates CEA's loss of free sunshine to stimulate plant photosynthesis. But especially in an environment of low borrowing costs to enable firms to invest in capital-intensive CEA methods, and in the face of increasing water scarcity that is more easily managed in compact spaces than in large, open fields, CEA is becoming increasingly viable as a means of expanding the supply of leafy greens and fast-growing (i.e., not tree-based) fruits.

The fourth group of innovations uses microbes and fungi to produce novel foods through a process broadly known as "precision fermentation." Fermentation is a centuries-old process used to make beer, cheese, etc., in virtually every culture globally. Recent advances in synthetic biology now enable labs to design micro-organisms (e.g., bacteria, microalgae, or yeasts) that produce more complex proteins from inexpensive feedstocks. This is the technology behind rapidly growing commercial enterprises such as Beyond Meat, Impossible Foods, and OmniFoods. This technology is not new; Quorn has employed the versatile mycoprotein since 1985 to make meat analogues. But precision fermentation has been taking off in the past few years as advances in (especially synthetic) biology have enabled cost reductions and improved customization of target proteins. In the first seven months of 2020 alone, these technologies attracted at least US$435 million in new investment, more than 3.5 times the capital raise by cultured/cellular meat companies globally (Shieber 2020). Precision fermentation methods can likely scale at costs below those of conventional systems for producing animal-source foods, generating a promising alternative to meet rapidly growing demand for more complex proteins without needing intermediation by livestock (Buckler and Rooney 2019; Tubb and Seba 2019).

As incomes increase, rapidly growing demand will inevitably deepen further for each of these de-agrarianized methods. Rising income, urbanization, and increased demand for shorter supply chains, and growing consumer concerns about nutrition, food safety, animal welfare, and the environmental impacts of conventional farm production methods will reinforce the momentum behind novel, land-saving food production methods, especially as companies and policymakers work to overcome consumers' natural skepticism about novel products

(Siegrist and Hartmann 2020). **The opportunity arises for technological leapfrogging in Africa and Asia**, in particular, as promising technologies that were previously unaffordable (e.g. CEA, precision fermentation) are becoming commercially viable at scale any place with reliable energy, adequate urban market size, and a literate workforce with sufficient basic scientific training. LMICs can use rural lands to farm carbon, solar, wind, and geothermal heat, not just crops and livestock, while simultaneously deploying novel technologies to design and deliver healthier foods—and remunerative urban and peri-urban jobs—based on shorter supply chains to meet growing urban food demand. In so doing, we can convert agri-food sectors from a GHG source to a sink, shift nutritional transitions in a healthier direction, and facilitate a structural transformation that harnesses looming demographic changes to simultaneously boost sustainability, resilience, inclusion, and healthy diets.

As promising as these land-saving methods are as a means to address sustainability, healthy diets, and resilience objectives simultaneously, they risk major social disruption, especially in rural areas that heavily depend on conventional farming. Major technological change inevitably unleashes what Joseph Schumpeter (1942) famously termed "creative destruction." **Without a concerted effort to transition rural economies, as lower costs of de-agrarianized food production increasingly undercut the profitability of conventional livestock and feed crop production, we run a real risk of a cascading calamity of farm bankruptcies, farmer suicides, and rural unrest**.

So what alternative sources of income exist for agricultural landowners and workers? We see at least three options. The first is renewable energy production, demand for which is growing rapidly around the world, especially as technological advances continue to drive down the costs of generating electricity from geothermal, solar, and wind sources and as off-grid alternatives have become increasingly viable. Lease royalties from energy companies and power utilities, and the non-farm value addition made feasible by reliable local power generation open up new livelihood options for agricultural communities. Indeed, there is reinforcing feedback between renewable energy production and novel, non-farm food production methods because cost-reducing technological change in one sector helps lower costs in the other. Relying just on unregulated energy markets and AVCs, therefore, seems a high-risk strategy for rural communities.

A second option is for governments to implement carbon taxes and invest more in establishing viable emissions trading systems (i.e., carbon markets) and the digital technologies necessary for low-cost, reliable verification of GHG fluxes to support monetizing sequestration activities. The current global average carbon price across both regulated and voluntary markets is only US\$2/tCO2, far below the US$40–80/tCO2 range necessary to cost-effectively reduce emissions in line with the Paris Agreement (HLCCP 2017; World Bank 2020). GHG sequestration is feasible in regenerative agriculture using sustainable farming practices, although concerns remain (Schlesinger and Amundson 2019). These environmental services can generate mitigation benefits to supplement agricultural earnings as farms diversify into harvesting GHG, solar, and wind, as well as commodities.

The third option is payments for ecosystem services (PES), which have grown popular worldwide, with an estimated US$40 billion or so in annual transactions (Salzman et al. 2018), with estimates for the potential revenues to the US' agriculture sector alone ranging as high as US$14 billion (Informa and IHS Markit 2018). PES have clearly demonstrated favorable impacts when well designed, although a range of design flaws continue to impede broad use and may limit sustainability gains (Jayachandran et al. 2017; Jack and Jayachandran 2019). Thus, PES are useful instruments, but no panacea. They appear to work most effectively in contexts involving few and large beneficiaries of the environmental services, such as hydroelectric companies or municipalities.

These alternative uses of agricultural lands create a terrific opportunity for policy innovation, in particular by repurposing farm subsidies. OECD (2020) estimates that across 54 countries which it tracks, transfers to the agricultural sector averaged US$708 billion/year for 2017–2019, of which fully US$425 billion was budgetary spending, with the rest coming through market-price support programs. Three-quarters of the amount goes to individual producers, mostly in forms that distort markets. Eliminating massive subsidies seems a political non-starter in most or all of the countries where they are large. But it may be politically feasible to transition from uncoupled farm payments or expensive market price supports to subsidies for farmer capital investments in renewable energy structures, in PES, in land conversion for GHG sequestration, and in the digital technologies—and supporting market infrastructure—necessary to monetize those energy and environmental services. A more forward-looking approach to the use of politically explosive farm subsidies can safeguard rural communities for the coming future when de-agrarianized production methods begin undercutting rural economies heavily dependent on conventional agricultural commodity production.

Facilitating land conversion from agriculture will also require action regarding land use rules. Secure land tenure is essential to induce investment in GHG mitigation in trees, soils, or cover crops, much less in installation of wind turbines or solar panel arrays. Concerted efforts will be necessary to overcome commonplace local opposition (e.g., "Not In My Backyard!" NIMBYism) regarding the siting of wind turbines, solar panels, protected areas for predators, etc. These are delicate processes but essential to transitioning rural landscapes.

Supply Chain Innovations

Purposeful changes are needed for AVCs that extend from the farm through to the consumer and end-of-life material considerations. We emphasize six key facets needing—and increasingly getting—attention from food and beverage companies, ingredient suppliers, global governance structures, non-governmental organizations, wholesale and retail operations, and national policy makers.

The first surrounds value chain certification standards. Many claims about a product's environmental, ethical, or healthful properties—its credence attributes—cannot be verified directly by purchase or consumption (Barrett 2021). This makes it difficult for firms to monetize the value of desirable product characteristics and, thus, to use market mechanisms to incentivize such innovations. In several regions, government agencies, like the US Food and Drug Administration and the European Food Safety Authority (EFSA), regulate heath claims based on scientific evidence on the label. Regarding environmental sustainability of products or services, a large European initiative is now evaluating a label called Product Environmental Footprint, which builds on prior prototypes and studies (e.g., Leach et al. 2016). Companies like Unilever propose to explicitly report associated GHG emissions on the packaging of tens of thousands of products. The International Organization for Standardization (ISO) standards for carbon labelling require a full life-cycle analysis and third-party verification, the cost of which poses a potential hurdle for small- and medium-sized companies and for mass labelling. A future area of innovation may focus on ways to reduce the cost of these assessments and potentially automate for large numbers of varied products. However, it also seems challenging to agree on a representative and simple sustainability indicator that consumers understand, that is widely adopted and recognized by different stakeholders, and that covers the various dimensions needed (Chaudhary et al. 2018; Chen et al. 2019). For example, nutrition and linked health impacts are essential, but are not considered in the Product Environmental Footprint.

We need accelerated convergence of food and ingredient supply chain certification schemes on key performance measures that catalyze the UN SDGs and an expanded set of Science Based Targets (SBTs).[15] Success in leap-frogging beyond the existing meta-system of certification standards will reflect four distinct refinements:

[15] SBTs are widely accepted targets voluntarily agreed by companies to set a clearly defined pathway towards medium- to longer-run goals. To date, these have focused almost exclusively on reducing GHG emissions so as to mitigate climate change. See https://sciencebasedtargets.org/ for further detail.

- unifying KPMs for social, economic, and environmental aspects;
- clarity and transparency for supply chain participants from consumer-to-the-farm around a single set of KPMs;
- a continuous improvement ecosystem of measures, protocols, resources, and consumer communication; and
- an easily adopted framework for governments to focus sustainable food system policy development and support structures.

The emergence of harmonized standards and associated measures, with traceable, trackable, scrapable product-level data, could ultimately supplant costly third-party certification if individual companies' and industries' compliance becomes fully transparent and independently verifiable by government regulators and consumer groups.

In the near term, KPMs within certification schemes need to evolve to reliably capture key indicators (discussed below) that directly support the SDGs and SBTs. In order to deliver broad-based change, in particular with small- and medium-sized value chain participants, certification schemes require reciprocity and KPM convergence so as not to unfairly burden upstream players, especially small-holder farmers (Loconto and Dankers 2014). Certifications are to be built around principles of continuous improvement rather than either achievement of a standard that is then passively maintained, or such a high entry hurdle that it dissuades parties from initiating the scheme (Blackman and Rivera 2011). KPMs must be supported through nonmanipulable tracking and traceability technologies of the sort we discuss below.

Certification frameworks would be best linked to relevant objectives and indicators. When a certification process is established, it brings an ecosystem of frameworks that support measurement, verification, transparency, capability building, and communication (e.g., third-party certification bodies, technical panels to oversee measures, standards and technical resources developed for user networks, etc.).

Designing for an ecosystem of measures, protocols, resources, and consumer communication acknowledges the ongoing infrastructure and support required to drive long-term continuous improvement across KPMs. Protocols and resources are the domain of value chain participants, certification bodies, auditors, civil society, and, where possible, government actors who all come together pre-competitively to build the elements of the scheme and a means of continuous improvement by establishing protocols, independent and verified auditing, best-practice sharing, training, and capability building.

Certification schemes create clear expectations about standards and compliance, thereby generating credibility and consumer trust at point of purchase. Such trust is essential to monetize latent consumer willingness to pay for credence attributes and thereby internalize key climate, environmental, and social externalities generated throughout the AVC. Standards must also be easily and reliably communicated to consumers in simple, easy-to-understand messaging and icons or logos that indicate verified performance and transparency. Avoidance of logo proliferation is important, however, so as not to sow confusion among consumers. Furthermore, while larger producers may be capable of achieving and maintaining multiple certifications, smallholders rarely can, so elimination of unique (and other high-cost) certification criteria and mutual recognition between platforms is essential to focus the value chain on clear outcomes that make a difference while meeting the needs of a diverse array of consumers and producers.

This requires more cooperation than presently occurs among AVC actors. It requires pre-competitive partnership of large-scale end users of food and ingredients with global governing bodies, relevant civil society organizations, existing certification bodies, suppliers, and implementation partners. While progressive enterprises should be encouraged to pilot innovative methods, in order to generate scalable, trusted methods, such experiments must be done in the spirit of shared learning to be incorporated into the meta-system to benefit all parties. This will also require a modular approach reflecting the heterogeneity of underlying AFSs and starting points.

The second key supply chain innovation space, closely related to certification, concerns consumer transparency. This has a robust foundation in food and beverage nutritional labeling that is currently coordinated through Codex Alimentarius, a collection of internationally adopted food standards and related texts jointly supervised by FAO and the World Health Organization (WHO). The Codex legacy of fact-based disclosure must be extended to key indicators that support the UN SDGs and SBTs, and HERS objectives more broadly (Box 6). The potential consumer and social benefits from food labeling are considerable but often limited by the imperfect information available to purchasers, along with consumer behavioral biases (Sunstein, in print).

Box 6: Towards fact-based sustainability labelling

In the early 1970s, consumer transparency in foods and beverages was improved with refinement of a standardized nutrition-facts table printed on pre-packaged foods. Its development was initially supported by the US Food and Drug Administration and has evolved to governance and technical oversight by Codex Alimentarius, a UN body under joint WHO and FAO direction (Codex Alimentarius 2017). Some version of the nutrition facts label is mandatory for packaged foods in 58 countries and voluntary in another 19 (EUFIC 2016). This adoption rate, with technical rigor and coordination through a central governance body, supports global consistency for package labels. The precise regulation of labelling compliance is carried out at the country level. This system supports consumer confidence in food and beverage nutritional disclosures, empowering consumers to make reliable inter-product assessments when making purchasing choices and enabling companies to elicit revenue from consumer valuation of improved nutritional content.

Nutrition Facts

8 servings per container

Serving size	**2/3 cup (55g)**
Amount per serving	
Calories	**230**
	% Daily Value*
Total Fat 8g	**10%**
Saturated Fat 1g	**5%**
Trans Fat 0g	
Cholesterol 0mg	**0%**
Sodium 160g	**7%**
Total Carbohydrate 37g	**13%**
Dietary Fiber 4g	**14%**
Total Sugars 12g	
Includes 10g Added Sugars	**20%**
Protein 3g	
Vitamin D 2mcg	10%
Calcium 260mg	20%
Iron 8mg	45%
Potassium 240mg	6%

*The % Daily Value (DV) tells you how much a nutrient in a serving of food contributes to a daily diet. 2,000 calories a day is used for general nutrition advice.

These elements—central governance, technical rigor based on agreed and credible measures, local enforcement, and transparent and fact-based disclosure focused at individual consumers—set a precedent relevant to the challenge of communicating other HERS-related product attributes to consumers at the point of purchase. Lessons from the nutrition-facts label experience can inform development and consumer-directed communications of sustainability key performance measures that support the SDGs and SBTs (Leach et al. 2016). Such labelling regimes can activate latent consumer valuation of product credence attributes, thereby internalizing spillover effects and generating revenues necessary to cover the costs of improving environmental, equity, and health outcomes associated with specific food products. With credible measures and certified quantification, food and beverage markets can compete on a more equal footing, transcending greenwashing concerns with enhanced transparency, benefiting a range of AVC participants, and thereby advancing fruitful product and process innovations.

Transparency must also include disclosure of provenance for ingredients that, when combined with nutritional and third-party verified credence characteristics, paves the way to consumer trust and informed decision making. Here again, emergent technologies to enable nonmanipulable tracking and traceability become important. A number of promising initiatives are in early stages, such as the European Union's Product Environmental Footprint pilots or Unilever's GHG labeling initiative, as described above. **The potential to create universally recognized and respected labels, backed by reliable standards and testing, that earn and maintain consumer and regulator trust opens up exciting opportunities** to induce beneficial innovation by incentivizing it for AVC firms.

The third class of promising supply chain innovations are in food processing and are based on different (1) operations (*structuring, conversion, stabilization*, and *separation*), (2) processes (physical: thermal, electro-magnetic, and mechanical; and biotechnological), or (3) product property scales (nano, micro, meso, and macro scale). Especially due to emerging needs for urban food production, small-scale modular factories (Mathys 2018) for processing close to production or urban environments (e.g., megacities) are receiving more attention, as improved energy and water delivery technologies and robotization reduce economies of scale. Focused process synthesis approaches (Westerberg 2004) to adapt new ingredients (e.g., plant-based ingredients in place of animal-sourced ones) and desired final product attributes, (e.g., preferred organoleptic properties) are especially crucial. These process-synthesis approaches can deliver innovative product property scales, from nano to macro, not only for mimicking meat structures, from myofibrils (meat fibers) to final structured product, but also for enabling emerging single-cell and plant-based protein-rich products with new structures and ingredients to deliver preferred organoleptic properties such as superior taste, nutrition, and mouth feeling.

Emerging *structuring/conversion* processes—such as advanced high-moisture extrusion, 3D printing, shear cell technology, spinning, and stem cell techniques (i.e. for lab meat)—enable innovative meat substitutes or new protein-rich products based on more sustainable proteins (Dekkers et al. 2018). New ways of food *stabilization*/preservation based on the Multi-Hurdle Technology (MHT) (Leistner and Gorris 1995) concept deliver safe food with higher qualities, including emerging physico-chemical hurdles to reduce water activity, such as solar driers combined with moisture control that allow smallholders to preserve fruits and vegetables; and physical hurdles with less thermal intensity, such as ultra-short thermal processes in milli seconds; high pressure (isostatic and dynamic), pulsed electric fields; low- or high-energy electron beam; or cold atmospheric pressure plasma processing (Reineke and Mathys 2020).

High throughput *separation* processes can clean/sanitize contaminated commodities (e.g. mycotoxin contaminated grains).[16] Building out the capability for precision fermentation or single cell biorefineries of lipids (e.g., polyunsaturated fatty acids), precision or cellular proteins, and carbohydrates (e.g., exopolysaccharides) with cascade-wise extraction of, first, functional and then, bulk ingredients will help to reduce AFSs' land and water footprint when done in ways that boost carbon sequestration and biodiversity. Multi-processing biorefineries will emerge to integrate various process innovations, much as already exist for grains, sugar, etc.

> WE EXPECT THAT PENALTIES AND INCENTIVES WILL BOTH BE NEEDED TO REMOVE PIGMENTS, ADDITIVES, AND POLYMERS THAT MAKE RECYCLING UNECONOMICAL CURRENTLY.

The fourth class of promising supply chain innovations concern packaging. Ultra-processed foods, in particular, are not only associated with adverse health outcomes, they also use extensive packaging that has serious disposal impacts

[16] Buhler provides a nice example: https://digital.buhlergroup.com/lumovision/.

worldwide, ranging from toxic compounds, to hazards to wildlife, to solid waste (Seferidi et al. 2020). Besides reduction of packaging materials, the transition from single-use plastics/virgin abiotic material to 100 percent recyclable, biodegradable, or compostable materials must quickly become the norm worldwide. This will require investment and legislation that supports sufficient recycling infrastructure (open loop, closed loop, and chemical) to match the packaging material being used and behaviors through the life cycle to recapture molecules for reuse. **We expect that penalties and incentives will both be needed to remove pigments, additives, and polymers that make recycling uneconomical currently**. Beyond enablement of recycling and renewable resource utilization, there remains a significant gap in available technologies via monomers and compostable organic packaging materials that feasibly deliver required barrier properties (oxygen, water vapor, light, aroma, etc.) or the technologies may not have the right physical properties (personal communication Prof. Selçuk Yildirim, ZHAW, Switzerland). Many activities in this space are running in industrial environments and are not published, hence the status quo is not quite clear. Recent developments in food-processing multinationals demonstrate the increasing focus on recyclable, biodegradable, or compostable packaging materials, for example, the 2019 establishment of the new Nestlé Institute of Packaging Sciences.

Rapid advances in waste management represent the fifth promising supply chain innovation space. In general, the waste management hierarchy indicates an order of preference for action to reduce and manage waste.[17] First comes prevention: preventing and reducing waste generation. Next comes reuse and preparation for reuse, giving the products a second life before they become waste. The next priority is recycling, consisting of any recovery operation by which waste materials are reprocessed into products, materials, or substances whether for the original or other purposes. This is followed by energy recovery, such as waste incineration that upgrades less inefficient incinerators. The lowest priority is disposal of waste, be it landfilling, incineration, pyrolysis, gasification, or other finalist solutions. This hierarchy is rapidly winning acceptance by local to national governments and is being incorporated into standard operating practices at successful companies (Hansen et al. 2002; UNEP 2013).

[17] According to FAO, Food loss is the decrease in the quantity or quality of food resulting from decisions and actions by food suppliers in the chain, excluding retailers, food service providers, and consumers. Food waste refers to the decrease in the quantity or quality of food resulting from decisions and actions by retailers, food service providers, and consumers.

Food waste and losses occur at different points in the value chain, each requiring different innovations. Many require behavioral change more than scientific or engineering advances. For example, according to the Rockefeller Foundation Report "ReFED: The Roadmap to Reduce US Food Waste," major impacts for food waste and loss reduction in the US are linked to awareness, traceability, and transparency (ReFED 2016). Food loss reduction strategies in low-income regions are complex and involve, for example, awareness-raising combined with training and organization of smallholders, and improved storage and preservation capacities (e.g. for fruits and vegetables), distribution, and logistics (Cattaneo et al. 2021). Awareness, traceability, and transparency are also needed here. A recent global assessment of nutritional and environmental losses embedded in food waste could serve as a base for tracking potential intervention impacts, supporting policies or investments, and engaging various stakeholders within the value chain (Chen et al. 2020).

Some further technical-focused solutions might include (1) distribution and storage of higher quality and fresher foods, stabilized/preserved by emerging MHT concepts at ambient temperatures instead of energetic and partially challenging cold chains; (2) building out uses and upcycling of AVC by-products, for example, providing ingredients for brewers, distillers, and manufacturers; or (3) technology for digitally customizing individual serving/portion sizes in away-from-home dining.

As mentioned above, the sustainability issues related to fertilizer use and soil depletion can be addressed by innovations at the nexus of sanitation, energy, and soil health. A range of possibilities are being explored to deal with organic byproducts of animal agriculture, industry, and human digestion. One set of options entails anerobic digester technologies that can be introduced into both private AVCs that are generating waste products that cannot be upcycled, and into urban settings that generate an estimated 2.8 billion metric tons of organic waste annually. Outputs from an anerobic digester can support local electricity production, and the solids can be combined with aqueous ammonia to produce an organic fertilizer for use by local farmers. The precision fermentation model would also contribute to more local production and less generated food waste.

The final promising innovation space in supply chains relates to initiatives to enhance value chain resilience to shocks. Some of those innovations are technological. For example, MHTs will support resiliency as they increase the ability to store and also reduce food waste. Others are less innovations than investments to reinforce or relocate key transport infrastructure. Sea level rise due to global warming poses an especially grave threat because seaports are overwhelmingly located in low-lying coastal zones and delta regions.[18] **Sea level rise will affect ports through incremental, as well as catastrophic, flooding that damages infrastructure and cargo.**[19] In 2005, Hurricane Katrina halted shipping at three Gulf ports in the US, which together handle 45 percent of the nation's agricultural goods and resulted in a 3 percent increase in food prices temporarily (Drabenstott and Henderson 2005). Recent studies reinforce the magnitude of sea level rises and the irreversible impacts of ice sheet loss on coastal populations and infrastructure (Garbe et al. 2020). Ports around the globe are under-prepared to cope with these challenges. A survey of seaports that collectively account for over 16 percent of global seaborne trade reveals that although 70 percent of the respondents have, or plan to have, emergency response measures, about 40 percent do not have, or do not plan to have, any vulnerability assessments, and 41 percent have yet to conduct any identification and evaluation of potential adaptation measures (Asariotis et al. 2018). The survey also reports that instead of soft adaptation strategies, such as changes in operations and management, respondent ports mainly chose hard engineering measures as the main strategy with an average cost of US$127.3 million.

The projected effects of sea level rise are quite spatially concentrated. Eight Asian countries—Bangladesh, China, India, Indonesia, Japan, Thailand, the Philippines, and Vietnam—are home to more than 70 percent of the world population now occupying land vulnerable to sea level rise (Kulp and Strauss 2019). Indonesia's recent decision to move its capital from a swelling and sinking coastal city, Jakarta, to eastern Borneo is partly a direct response to the perils posed by sea level rise. Bangladesh and Vietnam are especially vulnerable, as roughly one-third of each country's population will permanently fall below high tide line by 2100, even with a significant reduction in emissions. The most catastrophic cases will obviously be low-lying small island states, whose very existence may be imperiled by rising seas.

Singapore provides an illuminating example of how nations are adapting to various threats posed by AVC disruptions. Currently, Singapore produces only 10

[18] In addition, an estimated 80 airports worldwide could be underwater with the projected one-meter-sea level rise by 2100 under the IPCC (2019) business-as-usual scenario, including some of the busiest in the world, for example, Amsterdam Schiphol (Huang and Maghsadi 2020).

[19] Potential tidal modulation can also cause sedimentation, forcing expensive dredging in navigation channels and changes in operational timetables (Asariotis et al. 2018; Stenek et al. 2011; Nicholls et al. 2008; Admiraal 2011; Becker et al. 2013).

percent of the food its population consumes. Historically, this has worked fine, as inexpensive imports reliably supplied the island nation's food needs. Even before the COVID-19 pandemic threatened imports due to commercial freight shutdowns and export bans imposed by some exporters, thus making Singapore's low self-supply rate a vulnerability, the government was committed to substantially increasing its self-provisioning of food so as to reduce vulnerability to short-term disruptions arising from any of a host of shocks (Zulkifli 2020). The threats posed to Singapore's port infrastructure by sea level rise merely aggravate the looming problem. The nation has now made it a strategic priority to increase domestic production to satisfy 30 percent of its nutritional needs by 2030. This is fueling rapid upscaling of investments in CEA, circular feeds, and other forms of de-agrarianized food and feed production, given the scarcity of land on the island, as well as advances in food loss and waste recovery and in food processing so as to triple domestic supply within a decade.

Health and Nutrition Innovations

Important downstream innovations show particular promise in advancing the healthy diets objective, but that may also help advance other AFS goals. We coarsely lump these into three categories: new nutritious foods, nutritious supply chain innovations, and new frontiers in human nutrition.

New technologies are emerging to produce and formulate new nutritious foods or new variations of foods to ensure that the food supply is providing healthier foods while potentially, at the same time, addressing climate change and environmental concerns, as well as issues of equity and inclusion in food distribution. One such technology is 3D printing, which can make a three-dimensional object based on layer-by-layer deposition following computer aided design (Yang et al. 2017). With 3D printing, ingredients can be mixed and processed into intricate designs and shapes, introducing new flavors and textures that cannot be currently formulated by regular cooking processes. **3D printing has the potential to support personalized food manufacturing through home scale production**. Questions, nonetheless, remain about consumer acceptance of 3D printed foods. And it is unclear whether 3D printing would promote healthier diets or reduce food loss and waste.

Genetic modification (GM) of organisms is another technology that has grown through a range of advances in genetics and genomics that enable the change, removal, or addition of genes to crops and livestock that are believed beneficial for one reason or another. The earliest GM agri-food technologies promoted shelf-stability in tomatoes, stimulated lactation in cows, obviated the slaughter of calves in extracting rennet for cheesemaking, and especially introduced pest and/or herbicide resistance to field crops like canola, cotton, maize, and soy. These initial ventures were largely aimed at boosting or stabilizing production (Qaim 2016). **Second generation GM agri-food innovations increasingly address nutrition issues—such as micronutrient deficiencies—that remain prevalent in too many** LMICs (Glass and Fanzo 2017). To address micronutrient deficiencies, staple crops such as maize, rice, and wheat could either use GM technology or conventional or accelerated breeding to increase the nutritional content of vitamin A, zinc, or iron, for example, through an innovation known as biofortification (Bouis and Saltzman 2017; CAST 2020). One such example of a nutrient-rich GM crop is the controversial golden rice in which beta-carotene was built into the rice grain to produce a vitamin A—rich rice product (Regis 2019; Stokstad 2019).

The alternative proteins discussed above open up a range of prospective nutrition innovations. Precision protein (also known as single cell protein or microbial protein) is produced by a microbe (algae, fungi, yeast, or bacteria). The microbe may, or may not, be bioengineered, and the product may be secreted from the organism or processed within the cell. Cellular proteins (also known as cultured or tissue engineered meat) are produced as multi-cellular animal tissues that maintain cell structure through production. No matter the source method, plant proteins—which are almost always processed in some way—can be easily combined in various ways. The nutritional content of cultured meat may not be a significant concern because the nutritional composition of these foods can be modified, enriched, and fortified in the lab to match the foods found naturally (Sergelidis 2019). But challenges remain. Will consumers accept these novel foods (Bryant and Barnett 2019)? History shows that consumers are often suspicious of unnatural foods, at least initially (Chriki and Hocquette 2020). Other concerns include cost, taste, sustainability, and safety.

Reformulation is the process of altering a food or beverage product's processing or composition to improve the product's health profile or to reduce the content of harmful nutrients or ingredients (Scott et al. 2017). Reformulation encompasses both removing negative ingredients and nutrients, as well as adding positive ones to foods ranging from minimally to highly processed foods. Reformulation may be undertaken for reasons unrelated to better public health outcomes via improved nutrition. Companies can, and do, reformulate products for a variety of other reasons, including to increase nutrient density; to improve shelf-life, safety, and taste; to reduce costs; and to otherwise improve profitability (Box 7).

Box 7: Reformulation, fortification, and functionalization—incentivizing old innovations

There has been increased attention given to the health impacts of highly processed foods that are high in salt, added sugar, saturated and trans fats, and energy density, and low in fiber, protein, and micronutrients, and that also contribute to, and are associated with, overweight, obesity, and non-communicable diseases (Vandevijvere et al. 2019; Monteiro et al. 2013; Baker and Friel 2016; Baker et al. 2020; Hall, n.d.). Sub-optimal dietary outcomes have stimulated governmental nutrition policies to strive to reduce the intake of salt, added sugar, and unhealthy fats. Alongside promoting consumption of fresh nutritious foods (e.g., fruits, vegetables, and whole grains), the **reformulation, fortification, or functionalization** of processed foods may help improve diets in every food system. Can reduced processing of food—such as grinding, milling, and the removal of key nutrients—to promote more whole foods decrease the need to add back nutrients post-processing and reduce environmental footprints of the process overall (Seferidi et al. 2020)? The innovations in this space are less around food science than around aligning incentives.

Reformulation of foods can remove negative nutrients and/or add positive nutrients. Currently, it consists mainly of reducing the amount of salt, added sugar, saturated and trans fats, and the energy density in processed foods, largely to produce niche products to expand the range of consumer choice (Buttriss 2013). Reformulation can also increase healthy components, such as fiber, protein, micronutrients, or phytochemicals. **Fortification** adds essential vitamins and minerals to commonly consumed foods such as maize flour, edible oil, rice, salt, and wheat flour. It can also replace micronutrients lost during processing, such as with cereals, or address micronutrient deficiencies in the population, as with iodized salt (Das et al. 2019; Salam et al. 2019). **Functionalization** involves adding other beneficial ingredients that are specifically targeted to improve health (phytochemicals, pro-biotics, etc.).

While the main research focusing on reformulation, fortification, and/or functionalization concerns these processes' potential to improve nutrition and health, the main current industrial practices are for other, commercial purposes: decreasing costs, meeting changing consumer preferences, tapping into new consumer markets to boost sales or the company's public image, improving food safety and preservation, and/or complying with government regulations, where they exist).

Inducing more reformulation using existing technologies to promote healthy diets likely requires shifting incentives through labeling requirements, taxes, and/or regulatory constraints. Clear consumer signaling through labeling can incentivize companies to reformulate, particularly if labels carry warnings. Simple, easy-to-interpret front-of-pack labels that include stars, traffic lights, or other assessments of nutrition and health are increasingly effective and used in Chile, Australia, New Zealand, and the US, among others (Reyes et al. 2019; Chantal et al. 2017; Hersey et al. 2013; Jones et al. 2019). Companies reformulate products in order to avoid a low rating or a warning label (Vandevijvere and Vanderlee 2019).

So-called "sin taxes" are another tool. National and local taxes on sugar-sweetened beverages (SSB) and other energy-dense foods have been introduced in several countries (Hagenaars). In Mexico, Saudi Arabia, and Chile, SSB taxes were associated with an 8–24 percent reduction in purchases (Taillie et al. 2020). Sin taxes often face strenuous corporate resistance, however (Sainsbury et al. 2020).

National bans of certain ingredients (e.g., trans fats or salt) or requirements of nutrients in specific food vehicles (e.g., fortification of flours, oil, or salt) also shift industry incentives to adjust the product portfolio (Fanzo and McLaren 2020). Where private industry standards cannot converge around beneficial practices—as in the case of salt iodization in the US, for example—government regulatory standards may be necessary.

Ultimately, food industry actors need incentives—positive or negative—to reformulate foods not only in response to consumer preferences that can be manipulated through marketing, but equally, to improve consumer nutrition and health, as well as environmental sustainability in the face of the climate crisis. The food industry responds to new demands to make premium or superior products, as well as continual demands to make lower-cost products.

Once food moves through supply chains, innovations attempt to maintain or improve quality and ensure those foods are accessible and affordable. For at least 3 billion people, healthy foods remain unaffordable (Hirvonen et al. 2020; Headey and Alderman 2019; Bai et al. 2020; FAO 2020). The bitter irony is that affordable, healthy diets are especially inaccessible to the rural poor, who are most likely to work in AFSs. For example, 63–76 percent of India's rural poor could not afford a recommended diet in 2011 (Raghunathan et al. 2020). Globally, Bai et al. (2020) find that the minimum cost of a nutritious diet relative to household per capita expenditures falls with per capita income, access to electricity, and proximity to a city.

Food consumption can be influenced by ensuring nutritious foods are cheaper and unhealthy foods are more expensive (Eyles et al. 2012; Thow et al. 2014). **Taxes and subsidies can be used to shape prices and change dietary intake.** For example, taxes on SSBs can lead to a 20–50 percent reduction in consumption, while the subsidies for fruits and vegetables can lead to a 10–30 percent increase in consumption (Thow et al. 2014). However, taxes can be regressive, imposing greater economic burdens on the poor than on the wealthy. Combining taxes with healthy food subsidies—which have been far less common—could be one mechanism to mitigate the regressivity by allowing for populations to switch to healthier products without additional costs (Thow et al. 2010).

Nonmanipulable tracking and traceability using molecular markers, biomarkers, micro sensors, regulation, etc., can also improve food safety and help producers and intermediaries capture consumer valuation of foods' credence attributes. However, significant challenges and technical barriers must still be overcome. Food safety tracking and traceability systems are probably one of the best developed solutions in this domain in most high-income countries. However, much of the Global South still lacks sufficient safety tracking and traceability, with serious consequences, such as outbreaks of pathogenic bacteria or viruses, and chemical contaminations.

Block chain technology in agriculture and food supply chains has gained much attention recently. Is this a solution for nonmanipulable traceability? Significant challenges with this still-emerging technology exist around accessibility, governance, technical aspects (e.g., energy demands), policies, and regulatory frameworks (Kamilaris et al. 2019; Behnke and Janssen 2020). As with every innovation, we need to maximize the technology readiness level up to 9 (i.e., the actual system proven in an operational environment) before reaching strong conclusions, and we must learn from the ongoing innovation cycle.

Precision or personalized nutrition (PN) is an approach to addressing current nutrition problems using large quantities of detailed and multidimensional metabolic and health data to better understand the range of how human metabolism responds to diet. PN relies on a wide range of tools, including genomics, metabolomics, microbiomics, phenotyping, high-throughput analytical chemistry techniques, longitudinal tracking using body sensors, informatics, data science, and education and behavioral interventions to arrive at highly personalized and targeted dietary guidance and interventions (O'Sullivan et al. 2018).

Although many studies have been performed to identify genetic factors that explain the variability in metabolic response to specific diets, most findings are still relatively far from translatable for guidance. However, there are examples of findings that have already translated into guidance—including hypolactasia diagnosis, the ruling out of celiac disease, or phenylketonuria screening—which have led to tailored nutritional advice (avoiding lactose, gluten, and phenylalanine-containing products for at-risk individuals) based on genetics (de Toro-Martín et al. 2017).

Individualized approaches to PN remain expensive, though, and therefore may not be feasible in all settings due to budget constraints. Cost is a big reason why PN has thus far been targeted mainly to high-income environments, where individuals face very different nutritional challenges than do those in resource-poor settings. Moreover, **PN is not a substitute for public health infrastructure addressing underlying social, political, and economic inequities that are known drivers of population health outcomes**. There is much work to do in removing existing barriers (social, economic, political) to adequate diets. PN can fine tune once barriers are removed. Global populations may be diverse, so the call for diversified approaches to addressing diet-driven health problems makes sense on its face. But are individual differences in responses to diets really a significant driver of the global burden of diet-related disease? Thus far there is insufficient evidence that genotype-specific recommendations from direct-to-consumer genetic testing companies perform any better than "one-size-fits-all" recommendations (Loos 2019). For example, one recent study attempting to predict who would respond to dietary supplements of omega-3 fatty acids did not perform well out of sample (Marcotte et al. 2019). A second study examining the effects of dietary linoleic acid found an effect (Lankinen et al. 2019), but the magnitude was too small to be of use in precision nutrition (de Roos et al. 2019). While there is significant enthusiasm for PN-based methods, we do not yet see this as a high-potential area.

References

Admiraal, Jeroen. 2011. Case study: Vulnerability of the port of Rotterdam to climate change. In *Flood damage to port industry.* M.Sc. Thesis, Faculteit der Adarden Levenswtenschappen, Vrije Universiteit, Amsterdam.

Alderman, Harold, Ugo Gentilini, and Ruslan Yemtsov. 2017. *The 1.5 billion people question: Food, vouchers, or cash transfers?* Washington, DC: World Bank. https://doi.org/10.1596/978-1-4648-1087-9.

Altmann, Brianne A., Ruth Wigger, Marco Ciulu, and Daniel Mörlein. 2020. The effect of insect or microalga alternative protein feeds on broiler meat quality. *Journal of the Science of Food and Agriculture* 11: 4292. https://doi.org/10.1002/jsfa.10473.

Amelung, W., D. Bossio, W. de Vries, I. Kögel-Knabner, J. Lehmann, R. Amundson, R. Bol, C. Collins, R. Lal, J. Leifeld, B., and Minasny. 2020. Towards a global-scale soil climate mitigation strategy. *Nature Communications* 11 (1): 1–10.

Angelopoulou, Theodora, Athanasios Balafoutis, George Zalidis, and Dionysis Bochtis. 2020. From laboratory to proximal sensing spectroscopy for soil organic carbon estimation-A review. *Sustainability* (Switzerland) 12 (2): 1–24. https://doi.org/10.3390/su12020443.

Ariga, Joshua, Edward Mabaya, Michael Waithaka, and Maria Wanzala-Mlobela. 2019. Can improved agricultural technologies spur a green revolution in Africa? A multi-country analysis of seed and fertilizer delivery systems. *Agricultural Economics* 50 (1): 63–74.

Asariotis, Regina, Hassiba Benamara, and Viktoria Mohos-Naray. 2018. *Port industry survey on climate change impacts and adaptation*. UNCTAD Research Paper No. 18. https://unctad.org/system/files/official-document/ser-rp-2017d18_en.pdf.

Ba, Malick N., Ibrahim B. Baoua, Adama Kaboré, Laouali Amadou, Nassirou Oumarou, Clementine Dabire-Binso, and Antoine Sanon. 2014. Augmentative on-farm delivery methods for the parasitoid Habrobracon Hebetor Say (Hymenoptera: Braconidae) to control the millet head miner Heliocheilus Albipunctella (de Joannis) (Lepidoptera: Noctuidae) in Burkina Faso and Niger. *BioControl: Journal of the International Organization for Biological Control* 59 (6): 689. https://doi.org/10.1007/s10526-014-9613-8.

Bai, Yan, Robel Alemu, Steven A. Block, Derek Headey, and William A. Masters. 2020. Cost and affordability of nutritious diets at retail prices: Evidence from 177 countries. *Food Policy* (October): 101983. https://doi.org/10.1016/j.foodpol.2020.101983.

Bailey-Serres, Julia, Takeshi Fukao, Pamela Ronald, Abdelbagi Ismail, Sigrid Heuer, and David Mackill. 2010. Submergence tolerant rice: SUB1's journey from landrace to modern cultivar. *Rice* 3 (2): 138–147. https://doi.org/10.1007/s12284-010-9048-5.

Baker, Phillip, and Sharon Friel. 2016. Food systems transformations, ultra-processed food markets and the nutrition transition in Asia. *Globalization and Health* 12 (1): 80. https://doi.org/10.1186/s12992-016-0223-3.

Baker, Phillip, Priscila Machado, Thiago Santos, Katherine Sievert, Kathryn Backholer, Michalis Hadjikakou, Cherie Russell, et al. 2020. Ultra-processed foods and the nutrition transition: Global, regional and national trends, food systems transformations and political economy drivers. *Obesity Reviews: An Official Journal of the International Association for the Study of Obesity*, August. https://doi.org/10.1111/obr.13126.

Bänziger, Marianne, and Mark Cooper. 2001. Breeding for low input conditions and consequences for participatory plant breeding: Examples from tropical maize and wheat. *Euphytica* 122: 503–519. https://doi.org/10.1023/A:1017510928038.

Barbato, M., F. Hailer, M. Upadhyay, M. Del Corvo, L. Colli, R. Negrini, E.S. Kim, R.P. Crooijmans, T. Sonstegard, and P. Ajmone-Marsan. 2020. Adaptive introgression from indicine cattle into white cattle breeds from Central Italy. *Scientific Reports* 10 (1): 1–11. https://doi.org/10.1038/s41598-020-57880-4.

Barrett, Christopher B. (ed.). 2013. *Food security and sociopolitical stability*. Oxford: Oxford University Press.

Barrett, Christopher B. 2020. Benevolent patent extensions. Cornell University Working Paper.

Barrett, Christopher B. 2021. Overcoming global food security challenges through science and solidarity. *American Journal of Agricultural Economics* 103 (2): 422–447.

Barrett, Christopher B., Michael Carter, Jean-Paul Chavas, and Michael R. Carter (eds.). 2019. *The economics of poverty traps*. Chiciago: University of Chicago Press.

Barriga, Alicia, and Nathan Fiala. 2020. The supply chain for seed in Uganda: Where does it go wrong? *World Development* 130: 104928.

Barrios, Edmundo, Barbara Gemmill-Herren, Abram Bicksler, Emma Siliprandi, Ronnie Brathwaite, Soren Moller, Caterina Batello, and Pablo Tittonell. 2020. The 10 elements of agroecology: Enabling transitions towards sustainable agriculture and food systems through visual narratives. *Ecosystems and People* 16 (1): 230–247. https://doi.org/10.1080/26395916.2020.1808705.

Basso, Bruno, and John Antle. 2020. Digital agriculture to design sustainable agricultural systems. *Nature Sustainability* 3 (4): 254–256. https://doi.org/10.1038/s41893-020-0510-0.

Barroso, Fernando G., María-José. Sánchez-Muros, Macarena Segura, Elvira Morote, Alejandro Torres, Rebeca Ramos, and José-Luis Guil. 2017. Insects as food: Enrichment of larvae of Hermetia Illucens with omega 3 fatty acids by means of dietary modifications. *Journal of Food Composition and Analysis* 62 (September): 8–13. https://doi.org/10.1016/j.jfca.2017.04.008.

Bastagli, Francesca, Jessica Hagen-Zanker, Luke Harman, Valentina Barca, Georgina Sturge, Tanja Schmidt, and Luca Pellerano. 2016. Cash transfers: What does the evidence say? A rigorous review of programme impact and of the role of design and implementation features. https://doi.org/10.13140/RG.2.2.29336.39687.

Becker, Austin H., Michele Acciaro, Regina Asariotis, Edgard Cabrera, Laurent Cretegny, Philippe Crist, Miguel Esteban, et al. 2013. A note on climate change adaptation for seaports: A challenge for global ports, a challenge for global society. *Climatic Change* 120 (4): 683–695. https://doi.org/10.1007/s10584-013-0843-z.

Behnke, Kay, and M.F.W.H.A. Janssen. 2020. Boundary conditions for traceability in food supply chains using blockchain technology. *International Journal of Information Management* 52 (June): 101969. https://doi.org/10.1016/j.ijinfomgt.2019.05.025.

Behrman, Jere H., and John Hoddinott. 2005. *An evaluation of the impact of PROGRESA on pre-school child height*. Washington, DC: International Food Policy Research Institute.

Behrman, Jere R., and Emmanuel Skoufias. 2006. Mitigating myths about policy effectiveness: Evaluation of Mexico's antipoverty and human resource investment program. *Annals of the American Academy of Political and Social Science* 606: 244–275. https://doi.org/10.1177/0002716206288956.

Benachour, Nora, and Gilles-Eric Séralini. 2009. Glyphosate formulations induce apoptosis and necrosis in human umbilical, embryonic, and placental cells. *Chemical Research in Toxicology* 22 (1): 97–105. https://doi.org/10.1021/tx800218n.

Berendes, David M., Patricia J. Yang, Amanda Lai, Hu. David, and Joe Brown. 2018. Estimation of global recoverable human and animal faecal biomass. *Nature Sustainability* 1 (11): 679–685. https://doi.org/10.1038/s41893-018-0167-0.

Berti, Marisol, Russ Gesch, Christina Eynck, James Anderson, and Steven Cermak. 2016. Camelina uses, genetics, genomics, production, and management. *Industrial Crops and Products*. https://doi.org/10.1016/j.indcrop.2016.09.034.

Blackman, Alan, and Jorge Rivera. 2011. Producer-level benefits of sustainability certification. *Conservation Biology* 25 (6): 1176–1185. https://doi.org/10.1111/j.1523-1739.2011.01774.x.

Bloch, Sarah E., Min-Hyung Ryu, Bilge Ozaydin, and Richard Broglie. 2020. Harnessing atmospheric nitrogen for cereal crop production. *Current Opinion in Biotechnology* 62 (April): 181–188. https://doi.org/10.1016/j.copbio.2019.09.024.

Boffo, Riccardo, and Robert Patalano. 2020. *ESG investing: Practices, progress and challenges*. Paris: OECD. www.oecd.org/finance/ESG-Investing-Practices-Progress-and-Challenges.pdf.

Bonaudo, Thierry, Amaury Burlamaqui Bendahan, Rodolphe Sabatier, Julie Ryschawy, Stéphane Bellon, François Leger, Danièle Magda, and Muriel Tichit. 2014. Agroecological principles for the redesign of integrated crop–livestock systems. *European Journal of Agronomy, Integrated Crop-Livestock* 57 (July): 43–51. https://doi.org/10.1016/j.eja.2013.09.010.

Bouis, Howarth E., and Amy Saltzman. 2017. Improving nutrition through biofortification: A review of evidence from HarvestPlus, 2003 through 2016. *Global Food Security* 12 (March): 49–58. https://doi.org/10.1016/j.gfs.2017.01.009.

Bowles, Timothy M., Maria Mooshammer, Yvonne Socolar, Francisco Calderón, Michel A. Cavigelli, Steve W. Culman, William Deen, et al. 2020. Long-term evidence shows that crop-rotation diversification increases agricultural resilience to adverse growing conditions in North America. *One Earth* 2 (3): 284–293. https://doi.org/10.1016/j.oneear.2020.02.007.

Brookes, Graham, and Peter Barfoot. 2018. Environmental impacts of genetically modified (GM) crop use 1996–2016: Impacts on pesticide use and carbon emissions. *GM Crops & Food* 9 (3): 109–139. https://doi.org/10.1080/21645698.2018.1476792.

Bryant, Christopher J., and Julie C. Barnett. 2019. What's in a name? Consumer perceptions of in vitro meat under different names. *Appetite* 137 (June): 104–113. https://doi.org/10.1016/j.appet.2019.02.021.

Buckler, Edward, and Travis Rooney. 2019. Could a new generation of fermentation change the planet? Working Paper.

Bulak, Piotr, Cezary Polakowski, Katarzyna Nowak, Adam Waśko, Dariusz Wiącek, and Andrzej Bieganowski. 2018. Hermetia illucens as a new and promising species for use in entomoremediation. *Science of the Total Environment* 633 (August): 912–919. https://doi.org/10.1016/j.scitotenv.2018.03.252.

Buttriss, Judith L. 2013. Food reformulation: The challenges to the food industry. *The Proceedings of the Nutrition Society* 72 (1): 61–69. https://doi.org/10.1017/S0029665112002868.

Byrne, Jane. 2020. Report: Black soldier fly market will be worth US$2.57bn in 2030. FeedNavigator.com. Accessed November 7, 2020. https://www.feednavigator.com/Article/2020/01/22/Report-BSF-market-will-be-worth-US-2.57bn-in-2030.

Cabello, Felipe C., Henry P. Godfrey, Alexandra Tomova, Larisa Ivanova, Humberto Dolz, Ana Millanao, and Alejandro H. Buschmann. 2013. Antimicrobial use in aquaculture re-examined: Its relevance to antimicrobial resistance and to animal and human health. *Environmental Microbiology* 7: 1917. https://doi.org/10.1111/1462-2920.12134.

Cao, Ke, Joel Gehman, and Matthew G. Grimes. 2017. Standing out and fitting in: Charting the emergence of Certified B Corporations by industry and region. *Hybrid Ventures* 19 (1): 1–38. https://doi.org/10.1108/S1074-754020170000019001.

Capalbo, Susan M., John M. Antle, and Clark Seavert. 2017. Next generation data systems and knowledge products to support agricultural producers and science-based policy decision making. *Agricultural Systems* 155: 191–199. https://doi.org/10.1016/j.agsy.2016.10.009.

Caporgno, Martín P., and Alexander Mathys. 2018. Trends in microalgae incorporation into innovative food products with potential health benefits. *Frontiers in Nutrition* 5. https://doi.org/10.3389/fnut.2018.00058.

Cattaneo, Andrea, Marco V. Sánchez, Máximo Torero, and Rob Vos. 2021. Reducing food loss and waste: Five challenges for policy and research. *Food Policy* (September): 101974. https://doi.org/10.1016/j.foodpol.2020.101974.

Chantal, Julia, Serge Hercberg, World Health Organization, and Others. 2017. Development of a new front-of-pack nutrition label in France: The five-colour nutri-score. *Public Health Panorama* 3 (4): 712–725.

Charpentier, Myriam, and Giles Oldroyd. 2010. How close are we to nitrogen fixing cereals? *Current Opinion in Plant Biology* 13 (October): 556–564. https://doi.org/10.1016/j.pbi.2010.08.003.

Chaudhary, Abhishek, David Gustafson, and Alexander Mathys. 2018. Multi-indicator sustainability assessment of global food systems. *Nature Communications* 9 (1): 848. https://doi.org/10.1038/s41467-018-03308-7.

Chavas, Jean-Paul., Michael Aliber, and Thomas L. Cox. 1997. An analysis of the source and nature of technical change: The case of US agriculture. *Review of Economics and Statistics* 79 (3): 482–492.

Chen, Canxi, Abhishek Chaudhary, and Alexander Mathys. 2019. Dietary change scenarios and implications for environmental, nutrition, human health and economic dimensions of food sustainability. *Nutrients* 11 (4): 856. https://doi.org/10.3390/nu11040856.

Chen, Canxi, Abhishek Chaudhary, and Alexander Mathys. 2020. Nutritional and environmental losses embedded in global food waste. *Resources, Conservation and Recycling* 160. https://doi.org/10.1016/j.resconrec.2020.104912.

Chriki, Sghaier, and Jean-François Hocquette. 2020. The myth of cultured meat: A review. *Frontiers in Nutrition* 7 (February). https://doi.org/10.3389/fnut.2020.00007.

Chvátalová, V. 2019. A critical evaluation of EFSA's environmental risk assessment of genetically modified maize MON810 for honeybees and earthworms. *Environmental Sciences Europe* 31: 52. https://doi.org/10.1186/s12302-019-0238-5.

Codex Alimentarius. 2017. Guidelines on nutrition labelling. Report CAC/GL 2-1985, revised 2017. Geneva: FAO and WHO.

Cottrell, Richard S., Julia L. Blanchard, Benjamin S. Halpern, Marc Metian, and Halley E. Froehlich. 2020. Global adoption of novel aquaculture feeds could substantially reduce forage fish demand by 2030. *Nature Food* 1 (5): 301–308. https://doi.org/10.1038/s43016-020-0078-x.

Council for Agricultural Science and Technology (CAST). 2020. Food biofortification—Reaping the benefits of science to overcome hidden hunger. Issue Paper 69. Ames, IA: CAST.

Dalgaard, Tommy, Nicholas J. Hutchings, and John R. Porter. 2003. Agroecology, scaling and interdisciplinarity. *Agriculture, Ecosystems and Environment* 100: 39–51.

Das, Jai K., Rehana A. Salam, Salman Bin Mahmood, Anoosh Moin, Rohail Kumar, Kashif Mukhtar, Zohra S. Lassi, and Zulfiqar A. Bhutta. 2019. Food fortification with multiple

micronutrients: Impact on health outcomes in general population. *Cochrane Database of Systematic Reviews* 12 (December): CD011400.
DeFries, Ruth. 2018. *Trade-offs and synergies among climate resilience, human nutrition, and agricultural productivity of cereals—What are the implications for the agricultural research agenda?* CGIAR ISPC Science Forum. Stellenbosch, South Africa, October 10–12. https://www.scienceforum2018.org/sites/default/files/2018-09/SF18_background_paper_DeFries_0.pdf.
de Groot, Richard, Tia Palermo, Sudhanshu Handa, Luigi Peter Ragno, and Amber Peterman. 2017. Cash transfers and child nutrition: Pathways and impacts. *Development Policy Review* 35 (5): 621–643. https://doi.org/10.1111/dpr.12255.
de Roos, Baukje, Anna-Marja Aura, Maria Bronze, Aedin Cassidy, María-Teresa Garcia Conesa, Eileen R. Gibney, Arno Greyling, et al. 2019. Targeting the delivery of dietary plant bioactives to those who would benefit most: From science to practical applications. *European Journal of Nutrition* 58 (2): 65–73. https://doi.org/10.1007/s00394-019-02075-5.
De Silva, Sena S. 2012. Aquaculture: A newly emergent food production sector-and perspectives of its impacts on biodiversity and conservation. *Biodiversity and Conservation* 21 (12): 3187. https://doi.org/10.1007/s10531-012-0360-9.
Dekkers, Birgit L., Remko M. Boom, and Atze Jan van der Goot. 2018. Structuring processes for meat analogues. *Trends in Food Science & Technology* 81 (November): 25–36. https://doi.org/10.1016/j.tifs.2018.08.011.
Deutz, Andrew, Geoffrey M. Heal, Rose Niu, Eric Swanson, Terry Townshend, Li Zhu, Alejandro Delmar, Alqayam Meghji, Suresh A. Sethi, and John Tobin-de la Puente. 2020. *Financing nature: Closing the global biodiversity financing gap*. The Paulson Institute, The Nature Conservancy, and the Cornell Atkinson Center for Sustainability.
Drabenstott, Mark, and Jason Henderson. 2005. Katrina and Rita: Lingering effects on agriculture. *The Main Street Economist* [online]. Accessed October 12, 2020. http://www.kansascityfed.org/publicat/mse/MSE_1005.pdf.
Electris, Christi, Joshua Humphreys, Kristin Lang, David LeZaks and Jaime Silverstein. 2019. *Soil wealth: Investing in regenerative agriculture across asset classes*. Durham, NC: Croatan Institute. http://www.croataninstitute.org/soilwealth.
Ermakova, Maria, Florence R. Danila, Robert T. Furbank, and Susanne von Caemmerer. 2020. On the road to C4 rice: Advances and perspectives. *The Plant Journal* 101 (4): 940–950. https://doi.org/10.1111/tpj.14562.
Eskola, Mari, Gregor Kos, Christopher T. Elliott, Jana Hajšlová, J. Sultan Mayar, and Rudolf Krska. 2020. Worldwide contamination of food-crops with mycotoxins: Validity of the widely cited 'FAO estimate' of 25 percent. *Critical Reviews in Food Science and Nutrition* 60 (16): 1–17. https://doi.org/10.1080/10408398.2019.1658570.
European Food Information Council. 2016. Global update on nutrition labelling. Brussels: EUFIC. https://www.eufic.org/images/uploads/files/ExecutiveSummary.pdf.
Eyles, Helen, Cliona Ni Mhurchu, Nhung Nghiem, and Tony Blakely. 2012. Food pricing strategies, population diets, and non-communicable disease: A systematic review of simulation studies. *PLOS Medicine* 9 (12): e1001353. https://doi.org/10.1371/journal.pmed.1001353.
Fanzo, Jessica, Lawrence Haddad, Rebecca McLaren, Quinn Marshall, Claire Davis, Anna Herforth, Andrew Jones, et al. 2020. The food systems dashboard is a new tool to

inform better food policy. *Nature Food* 1 (5): 243–246. https://doi.org/10.1038/s43016-020-0077-y.

Fanzo, Jessica, and Rebecca McLaren. 2020. *The product reformulation journey so far: An assessment*. Geneva, Switzerland: GAIN Discussion Paper Series 8. https://doi.org/10.36072/dp.8.

FAO, IFAD, UNICEF, WFP and WHO. 2020. *The state of food security and nutrition in the world 2020: Transforming food systems for affordable healthy diets*. Rome, FAO.

Finckh, Maria R. 2008. Integration of breeding and technology into diversification strategies for disease control in modern agriculture. *European Journal of Plant Pathology* 121 (3): 399–409. https://doi.org/10.1007/s10658-008-9273-6.

Friede, Gunnar, Timo Busch, and Alexander Bassen. 2015. ESG and financial performance: Aggregated evidence from more than 2000 empirical studies. *Journal of Sustainable Finance & Investment* 5 (4): 210–233. https://doi.org/10.1080/20430795.2015.1118917.

Fry, Jillian P., David C. Love, Graham K. MacDonald, Paul C. West, Peder M. Engstrom, Keeve E. Nachman, and Robert S. Lawrence. 2016. Environmental health impacts of feeding crops to farmed fish. *Environment International* 91: 201–214. https://doi.org/10.1016/j.envint.2016.02.022.

Fuglie, Keith. 2016. The growing role of the private sector in agricultural research and development world-wide. *Global Food Security* 10 (September): 29–38. https://doi.org/10.1016/j.gfs.2016.07.005.

Galloway, James N., Allison M. Leach, Albert Bleeker, and Jan Willem Erisman. 2013. A chronology of human understanding of the nitrogen cycle. *Philosophical Transactions of the Royal Society B: Biological Sciences* 368 (1621): 20130120. https://doi.org/10.1098/rstb.2013.0120.

Garbe, Julius, Torsten Albrect, Anders Levermann, Jonathan F. Donges, and Ricarda Winkelmann. 2020. The hysteresis of the Antarctic Ice Sheet. *Nature* 585 (September): 538–544. https://doi.org/10.1038/s41586-020-2727-5.

Gawande, Atul. 2013. Slow ideas. *The New Yorker* (July): 1–15.

Gentilini, Ugo, Mohamed Almenfi, Pamela Dale, Robert Palacios, Harish Natarajan, Guillermo Alfonso Galicia Rabadan, Yuko Okamura, John Blomquist, Miglena Abels, Gustavo Demarco, and Indhira Santos. 2020a. *Social protection and jobs responses to COVID-19: A real-time review of country measures*. Washington, DC: World Bank. https://openknowledge.worldbank.org/handle/10986/33635.

Gentilini, Ugo, Margaret, Grosh, Jamele, Rigolini, Ruslan, Yemtsov. 2020b. *Exploring universal basic income: A guide to navigating concepts, evidence, and practices*. Washington, DC: World Bank. https://openknowledge.worldbank.org/handle/10986/32677.

Gentry, Rebecca R., Halley E. Froehlich, Dietmar Grimm, Peter Kareiva, Michael Parke, Michael Rust, Steven D. Gaines, and Benjamin S. Halpern. 2017. Mapping the global potential for marine aquaculture. *Nature Ecology & Evolution* 1 (9): 1317–1324. https://doi.org/10.1038/s41559-017-0257-9.

Ghislain, Marc, Arinaitwe Abel Byarugaba, Eric Magembe, Anne Njoroge, Cristina Rivera, María Lupe Román, José Carlos Tovar, et al. 2019. Stacking three late blight resistance genes from wild species directly into African Highland potato varieties confers complete field resistance to local blight races. *Plant Biotechnology Journal* 17 (6): 1119–1129. https://doi.org/10.1111/pbi.13042.

Glass, Sara, and Jessica Fanzo. 2017. Genetic modification technology for nutrition and improving diets: An ethical perspective. *Current Opinion in Biotechnology* 44: 46–51. https://doi.org/10.1016/j.copbio.2016.11.005.

Gliessman, Stephen R. 2016. Transforming food systems with agroecology. *Agroecology and Sustainable Food Systems* 40 (3): 187–189. https://doi.org/10.1080/21683565.2015.1130765.

Global Impact Investing Network (GIIN). 2020. *Annual impact investor survey 2020*. https://thegiin.org/assets/GIINpercent20Annualpercent20Impactpercent20Investorpercent20Surveypercent202020.pdf.

Godfray, H. Charles J. 2015. The debate over sustainable intensification. *Food Security* 7: 199–208. https://doi.org/10.1007/s12571-015-0424-2.

Gold, Moritz, Jeffery K. Tomberlin, Stefan Diener, Christian Zurbrügg, and Alexander Mathys. 2018. Decomposition of biowaste macronutrients, microbes, and chemicals in black soldier fly larval treatment: A review. *Waste Management* 82 (December): 302–318. https://doi.org/10.1016/j.wasman.2018.10.022.

Gonsalves, D. 1998. Control of papaya ringspot virus in papaya: A study. *Annual Review of Phytopathology* 36: 415–437. https://doi.org/10.1146/annurev.phyto.36.1.415.

González, F.G., N. Rigalli, P.V. Miranda, M. Romagnoli, K.F. Ribichich, F. Trucco, et al. 2020. An interdisciplinary approach to study the performance of second-generation genetically modified crops in field trials: A case study with soybean and wheat carrying the sunflower HaHB4 transcription factor. *Frontiers in Plant Science* 11 (March): 1–15. https://doi.org/10.3389/fpls.2020.00178.

Gould, Fred, Zachary S. Brown, and Jennifer Kuzma. 2018. Wicked evolution: Can we address the sociobiological dilemma of pesticide resistance? *Science* 360 (6390): 728–732. https://doi.org/10.1126/science.aar3780.

Graff, Gregory D., Felipe de Figueiredo Silva, and David Zilberman. 2020. Venture capital and the transformation of private R&D for agriculture. NBER Chapters. In *Economics of Research and Innovation in Agriculture*. National Bureau of Economic Research, Inc.

Griscom, Bronson W., Justin Adams, Peter W. Ellis, Richard A. Houghton, Guy Lomax, Daniela A. Miteva, William H. Schlesinger, et al. 2017. Natural climate solutions. *Proceedings of the National Academy of Sciences* 114 (44): 11645–11650.

Grossi, C. E. M., E. Fantino, F. Serral, M. S. Zawoznik, D. A. Fernandez Do Porto, and R. M. Ulloa. 2020. *Methylobacterium* sp. 2A is a plant growth-promoting rhizobacteria that has the potential to improve potato crop yield under adverse conditions. *Frontiers in Plant Science* 11 (February): 1–15. https://doi.org/10.3389/fpls.2020.00071.

Hall, Kevin D. n.d. Ultra-processed diets cause excess calorie intake and weight gain: A one-month inpatient randomized controlled trial of ad libitum food intake. https://doi.org/10.31232/osf.io/w3zh2.

Hansen, Wenke; Maria Christopher, and Maic Verbuecheln. 2002. *EU waste policies and challenges for local and regional authorities*.

Harder, Robin, Rosanne Wielemaker, Tove A. Larsen, Grietje Zeeman, and Gunilla Öberg. 2019. Recycling nutrients contained in human excreta to agriculture: Pathways, processes, and products. *Critical Reviews in Environmental Science and Technology* 49 (8): 695–743. https://doi.org/10.1080/10643389.2018.1558889.

Hawkes, Corinna, Marie T. Ruel, Leah Salm, Bryony Sinclair, and Francesco Branca. 2020. Double-duty actions: Seizing programme and policy opportunities to address malnutrition in all its forms. *The Lancet* 395 (10218): 142–155. https://doi.org/10.1016/S0140-6736(19)32506-1.

Headey, Derek D., and Harold H. Alderman. 2019. The relative caloric prices of healthy and unhealthy foods differ systematically across income levels and continents. *Journal of Nutrition* 149 (11): 2020–2033. https://doi.org/10.1093/jn/nxz158.

Henchion, Maeve, Maria Hayes, Anne Maria Mullen, Mark Fenelon, and Brijesh Tiwari. 2017. Future protein supply and demand: Strategies and factors influencing a sustainable equilibrium. *Foods* 6 (7): 53.

Heredia, Lubasha, Simon Bartletta, Joe Carrubba, Dean Frankle, Katsuyoshi Kurihara, Benoît Macé, Edoardo Palmisani, et al. 2020. *Global asset management 2020: Protect, adapt, and innovate*. Boston: Boston Consulting Group. https://www.bcg.com/en-us/publications/2020/global-asset-management-protect-adapt-innovate.

Herren, Hans R., and Peter Neuenschwander. 1991. Biological control of cassava pests in Africa. *Annual Review of Entomology*. https://doi.org/10.1146/annurev.en.36.010191.001353.

Herrero, Mario, Petr Havlík, Hugo Valin, An. Notenbaert, Mariana C. Rufino, Philip K. Thornton, Michael Blümmel, Franz Weiss, Delia Grace, and Michael Obersteiner. 2013. Biomass use, production, feed efficiencies, and greenhouse gas emissions from global livestock systems. *Proceedings of the National Academy of Sciences* 110 (52): 20888–20893. https://doi.org/10.1073/pnas.1308149110.

Herrero, Mario, Philip K. Thornton, Daniel Mason-D'Croz, Jeda Palmer, Tim G. Benton, Benjamin L. Bodirsky, Jessica R. Bogard, et al. 2020. Innovation can accelerate the transition towards a sustainable food system. *Nature Food* 1 (5): 266–272. https://doi.org/10.1038/s43016-020-0074-1.

Herrero, Mario, Philip K. Thornton, Daniel Mason-D'Croz, J. Palmer, B. L. Bodirsky, P. Pradhan, C. B. Barrett, T. G. Benton, et al. 2021. Articulating the impact of food systems innovation on the Sustainable Development Goals. *The Lancet Planetary Health* 5 (1): e50–e62. https://doi.org/10.1016/S2542-5196(20)30277-1.

Hersey, James C., Kelly C. Wohlgenant, Joanne E. Arsenault, Katherine M. Kosa, and Mary K. Muth. 2013. Effects of front-of-package and shelf nutrition labeling systems on consumers. *Nutrition Reviews* 71 (1): 1–14. https://doi.org/10.1111/nure.12000.

Hidrobo, Melissa, John Hoddinott, Neha Kumar, and Meghan Olivier. 2018. Social protection, food security, and asset formation. *World Development* 101 (January): 88–103. https://doi.org/10.1016/j.worlddev.2017.08.014.

High-Level Commission on Carbon Prices (HLCCP). 2017. *Report of the high-level commission on carbon prices*. Washington, DC: World Bank.

High Level Panel of Experts on Food Security and Nutrition (HLPE). 2019. *Agroecological and other innovative approaches for sustainable agriculture and food systems that enhance food security and nutrition*. Rome.

Hirvonen, Kalle, Yan Bai, Derek Headey, and William A. Masters. 2020. Affordability of the EAT–Lancet reference diet: A global analysis. *The Lancet Global Health* 8 (1): e59–e66. https://doi.org/10.1016/S2214-109X(19)30447-4.

Huang, Tina, and Noah Maghsadi. 2020. Runways underwater: Maps show where rising seas threaten 80 airports around the world. *Resource Watch*. Blog post. https://blog.resourcewatch.org/2020/02/05/runways-underwater-maps-show-where-rising-seas-threaten-80-airports-around-the-world/.

International Labour Organization (ILO). 2020. *Social protection spotlight: Financing gaps in social protection*. Geneva: ILO.

IPBES. 2019. Summary for policymakers of the global assessment report on biodiversity and ecosystem services of the Intergovernmental Science-Policy Platform on Biodiversity and Ecosystem Services. S. Díaz, J. Settele, E.S. Brondízio, H.T. Ngo, M. Guèze, J. Agard, A. Arneth, P. Balvanera, K.A. Brauman, S.H.M. Butchart, K.M.A. Chan, L.A. Garibaldi, K. Ichii, J. Liu, S.M. Subramanian, G.F. Midgley, P. Miloslavich, Z. Molnár, D. Obura, A. Pfaff, S. Polasky, A. Purvis, J. Razzaque, B. Reyers, R. Roy Chowdhury, Y.J. Shin, I.J. Visseren-Hamakers, K.J. Willis, and C.N. Zayas, eds. IPBES secretariat, Bonn, Germany, 56 pages. https://doi.org/10.5281/zenodo.3553579.

IPCC. 2019. Climate change and land: An IPCC special report on climate change, desertification, land degradation, sustainable land management, food security, and greenhouse gas fluxes in terrestrial ecosystems. P.R. Shukla, J. Skea, E. Calvo Buendia, V. Masson-Delmotte, H.-O. Pörtner, D.C. Roberts, P. Zhai, R. Slade, S. Connors, R. van Diemen, M. Ferrat, E. Haughey, S. Luz, S. Neogi, M. Pathak, J. Petzold, J. Portugal Pereira, P. Vyas, E. Huntley, K. Kissick, M. Belkacemi, J. Malley, eds. https://www.ipcc.ch/srccl/.

IPES-Food. 2016. *From uniformity to diversity: A paradigm shift from industrial agriculture to diversified agroecological systems*. International Panel of Experts on Sustainable Food systems.

IPES-Food. 2018. *Breaking away from industrial food and farming systems: Seven case studies of agroecological transition*. International Panel of Experts on Sustainable Food systems.

Informa and IHS Markit. 2018. *Economic assessment for ecosystem service market credits from agricultural working lands*. https://ecosystemservicesmarket.org/wp-content/uploads/2019/09/Informa-IHS-Markit-ESM-Study-Sep-19.pdf.

International Platform on Sustainable Finance (IPSF). 2020. Annual report. Washington, DC: Financial Stability, Financial Services and Capital Markets Union.

ISAAA. 2019. *ISAAA in 2019: Accomplishment report*. Ithaca: ISAAA.

Jack, B. Kelsey., and Seema Jayachandran. 2019. Self-selection into payments for ecosystem services programs. *Proceedings of the National Academy of Sciences* 116 (12): 5326–5333. https://doi.org/10.1073/pnas.1802868115.

Jaganathan, Deepa, Karthikeyan Ramasamy, Gothandapani Sellamuthu, Shilpa Jayabalan, and Gayatri Venkataraman. 2018. CRISPR for crop improvement: An update review. *Frontiers in Plant Science* 9 (July): 1–17. https://doi.org/10.3389/fpls.2018.00985.

Jayachandran, Seema, Joost de Laat, Eric F. Lambin, Charlotte Y. Stanton, Robin Audy, and Nancy E. Thomas. 2017. Cash for carbon: A randomized trial of payments for ecosystem services to reduce deforestation. *Science* 357 (6348): 267–273. https://doi.org/10.1126/science.aan0568.

Jean, Neal, Marshall Burke, W. Michael Xie, Matthew Davis, David B. Lobell, and Stefano Ermon. 2016. Combining satellite imagery and machine learning to predict poverty. *Science* 353 (6301): 790–794. https://doi.org/10.1126/science.aaf7894.

Jensen, Nathaniel D., and Christopher B. Barrett. 2016. Agricultural index insurance for development. *Applied Economic Perspectives and Policy* 39 (2): 199–219.

Jiao, Xiaoqiang, Yang Lyu, Xiaobin Wu, Haigang Li, Lingyun Cheng, Chaochun Zhang, Lixing Yuan, et al. 2016. Grain production versus resource and environmental costs: Towards increasing sustainability of nutrient use in China. *Journal of Experimental Botany* 67 (17): 4935–4949. https://doi.org/10.1093/jxb/erw282.

Jiao, Xiao-qiang, Hong-yan Zhang, Wen-qi Ma, Chong Wang, Xioa-lin Li, and Fusuo Zhang. 2019. Science and Technology Backyard: A novel approach to empower smallholder farmers for sustainable intensification of agriculture in China. *Journal of Integrative Agriculture* 18 (8): 1657–1666. https://doi.org/10.1016/S2095-3119(19)62592-X.

Jiao, Xiaoqiang, Derara S. Feyisa, Jasper Kanomanyanga, D. Ngula Muttendango, Mudare Shingirai, Amadou Ndiaye, Bilisuma Kabeto, Felix Dapare, and Fusuo Zhang. 2020. Science and Technology Backyard model: Implications for sustainable agriculture in Africa. *Frontiers of Agricultural Science and Engineering*. https://doi.org/10.15302/J-FASE-2020360.

Jones, Andrew D. 2017a. Critical review of the emerging research evidence on agricultural biodiversity, diet diversity, and nutritional status in low-and middle-income countries. *Nutrition Reviews* 75 (10): 769–782. https://doi.org/10.1093/nutrit/nux040.

Jones, Andrew D. 2017b. On-farm crop species richness is associated with household diet diversity and quality in subsistence-and market-oriented farming households in Malawi. *The Journal of Nutrition* 147 (1): 86–96. https://doi.org/10.3945/jn.116.235879.

Jones, Alexandra, Bruce Neal, Belinda Reeve, Cliona Ni Mhurchu, and Anne Marie Thow. 2019. Front-of-pack nutrition labelling to promote healthier diets: Current practice and opportunities to strengthen regulation worldwide. *BMJ Global Health* 4 (6). https://doi.org/10.1136/bmjgh-2019-001882.

Jones, Shawn W., Alon Karpol, Sivan Friedman, Biniam T. Maru, and Bryan P. Tracy. 2020. Recent advances in single cell protein use as a feed ingredient in aquaculture. *Current Opinion in Biotechnology* 61 (February): 189–197. https://doi.org/10.1016/j.copbio.2019.12.026.

Kamilaris, Andreas, Agusti Fonts, and Francesc X. Prenafeta-Boldύ. 2019. The rise of blockchain technology in agriculture and food supply chains. *Trends in Food Science & Technology* 91 (September): 640–652. https://doi.org/10.1016/j.tifs.2019.07.034.

King, K.C., and C.M. Lively. 2012. Does genetic diversity limit disease spread in natural host populations? *Heredity* 109 (4): 199–203. https://doi.org/10.1038/hdy.2012.33.

Krebs, Julius, and Sonja Bach. 2018. Permaculture-scientific evidence of principles for the agroecological design of farming systems. *Sustainability* 10 (9): 1–24.

Kremer, Michael, Jonathan Levin, and Christopher M. Snyder. 2020. Advance market commitments: Insights from theory and experience. *AEA Papers and Proceedings* 110 (May): 269–273. https://doi.org/10.1257/pandp.20201017.

Kronebusch, Natalie, and Amy Damon. 2019. The impact of conditional cash transfers on nutrition outcomes: Experimental evidence from Mexico. *Economics and Human Biology* 33: 169–180. https://doi.org/10.1016/j.ehb.2019.01.008.

Kulp, Scott A., and Benjamin H. Strauss. 2019. New elevation data triple estimates of global vulnerability to sea-level rise and coastal flooding. *Nature Communications* 10 (1): 1–12. https://doi.org/10.1038/s41467-019-12808-z.

Laborde, David, Marie Parent, and Carin Smaller. 2020. *Ending hunger, increasing incomes, and protecting the climate: What would it cost donors?* Washington, DC: Ceres2030, International Institute for Sustainable Development, and International Food Policy Research Institute.

Ladha, J.K., M.L. Jat, Claire M. Stirling, Debashis Chakraborty, Prajal Pradhan, Timothy J. Krupnik, Tek B. Sapkota, et al. 2020. Achieving the Sustainable Development Goals in agriculture: The crucial role of nitrogen in cereal-based systems. *Advances in Agronomy* 163: 39–116. https://doi.org/10.1016/bs.agron.2020.05.006.

Lal, Rattan. 2016. Soil health and carbon management. *Food and Energy Security* 4: 212. https://doi.org/10.1002/fes3.96.

Lankinen, Maria A., Alexander Fauland, Bun-Ichi Shimizu, Jyrki Ågren, Craig E. Wheelock, Markku Laakso, Ursula Schwab, and Jussi Pihlajamäki. 2019. Inflammatory response to dietary linoleic acid depends on FADS1 genotype. *The American Journal of Clinical Nutrition* 109 (1): 165–175. https://doi.org/10.1093/ajcn/nqy287.

Leach, Allison M., Kyle A. Emery, Jessica Gephart, Kyle F. Davis, Jan Willem Erisman, Adrian Leip, Michael L. Pace, et al. 2016. Environmental impact food labels combining carbon, nitrogen, and water footprints. *Food Policy* 61: 213–223.

Lednev, Georgiy, Maxim Levchenko, and Igor Kazartsev. 2020. Entomopathogenic microorganisms in locusts and grasshoppers populations and prospects for their use for control of this pest group. *BIO Web of Conferences* 21 (January): 00025. https://doi.org/10.1051/bioconf/20202100025.

Leistner, Lothar, and Leon G. M. Gorris. 1995. Food preservation by hurdle technology. *Trends in Food Science & Technology* 6 (2): 41–46. https://doi.org/10.1016/S0924-2244(00)88941-4.

Lemmon, Zachary H., Nathan T. Reem, Justin Dalrymple, Sebastian Soyk, Kerry E. Swartwood, Daniel Rodriguez-Leal, et al. 2018. Rapid improvement of domestication traits in an orphan crop by genome editing. *Nature Plants*. https://doi.org/10.1038/s41477-018-0259-x.

Li, H.P., S.G. Zhang, X.Q. Rao, X.L. Ruan, G.H. Zhou, and H.Z. Fan. 2007. Safety evaluation of transgenic papaya 'Huanong No. 1' resistant to PRSV. Proceedings of the annual meeting of the Chinese Society for plant pathology.

Li, Chunjie, Ellis Hoffland, Thomas W. Kuyper, Yang Yu, Chaochun Zhang, Haigang Li, Fusuo Zhang, and Wopke van der Werf. 2020a. Syndromes of production in intercropping impact yield gains. *Nature Plants*: 1–8. https://doi.org/10.1038/s41477-020-0680-9.

Li, Chunjie, Ellis Hoffland, Thomas W. Kuyper, Yang Yu, Haigang Li, Chaochun Zhang, Fusuo Zhang, and Wopke van der Werf. 2020b. Yield gain, complementarity and competitive dominance in intercropping in China: A meta-analysis of drivers of yield gain using additive partitioning. *European Journal of Agronomy* 113 (February): 125987. https://doi.org/10.1016/j.eja.2019.125987.

Lobell, David B., George Azzari, Marshall Burke, Sydney Gourlay, Zhenong Jin, Talip Kilic, and Siobhan Murray. 2020. Eyes in the sky, boots on the ground: Assessing satellite- and ground-based approaches to crop yield measurement and analysis. *American Journal of Agricultural Economics* 102 (1): 202–219. https://doi.org/10.1093/ajae/aaz051.

Loconto, A., and C. Dankers. 2014. Voluntary standards: Impacting smallholders' market participation. Impact of international voluntary standards on smallholder market participation in developing countries. Food and Agriculture Organization of the United Nations, Rome.

Loos, Ruth J.F. 2019. From nutrigenomics to personalizing diets: Are we ready for precision medicine? *The American Journal of Clinical Nutrition* 109 (1): 1–2. https://doi.org/10.1093/ajcn/nqy364.

Lyu, Y., H.L. Tang, H.B. Li, F.S. Zhang, Z. Rengel, W.R. Whalley, and J.B. Shen. 2016. Major crop species show differential balance between root morphological and physiological responses to variable phosphorus supply. *Frontiers in Plant Science* 7: 1939. https://doi.org/10.3389/fpls.2016.01939.

Magonziwa, Blessing, Steven Vanek, John Ojiem, and Steven Fonte. 2020. A soil tool kit to evaluate soil properties and monitor soil health changes in smallholder farming contexts. *Geoderma* 376: 114539. https://doi.org/10.1016/j.geoderma.2020.114539.

Manley, James, Seth Gitter, and Vanya Slavchevska. 2012. *How effective are cash transfer programmes at improving nutritional status? A rapid evidence assessment of programmes' effects on anthropometric outcomes*. London: EPPI-Centre, Social Science Research Unit, Institute of Education, University of London.

Mao, Yanfei, Jose R. Botella, Yaoguang Liu, Jian-Kang Zhu. 2019. Gene editing in plants: Progress and challenges. *National Science Review* 6 (3): 421–437. https://doi.org/10.1093/nsr/nwz005.

Martin, Guillaume, Marc Moraine, Julie Ryschawy, Marie-Angélina Magne, Masayasu Asai, Jean-Pierre Sarthou, Michel Duru, and Olivier Therond. 2016. Crop–livestock integration beyond the farm level: A review. *Agronomy for Sustainable Development* 36 (3): 53. https://doi.org/10.1007/s13593-016-0390-x.

Mason-D'Croz, Daniel, Timothy B. Sulser, Keith Wiebe, Mark W. Rosegrant, Sarah K. Lowder, Alejandro Nin-Pratt, Dirk Willenbockel, et al. 2019. Agricultural investments and hunger in Africa modeling potential contributions to SDG 2 – zero hunger. *World Development* 116 (April): 38–53. https://doi.org/10.1016/j.worlddev.2018.12.006.

Masuda, Yuta, Teevrat Garg, Ike Anggraeni, Nicholas H. Wolff, Kristie L. Ebi, Edward (Eddie) Game, Jennifer Krenz, and June Spector. 2020. Heat exposure from tropical deforestation decreases cognitive performance of rural workers: An experimental study. *Environmental Research Letters*https://doi.org/10.1088/1748-9326/abb96c

Mathys, Alexander. 2018. Perspective of micro process engineering for thermal food treatment. *Frontiers in Nutrition* 5.https://doi.org/10.3389/fnut.2018.00024.

McCauley, Janice I., Leen Labeeuw, Ana C. Jaramillo-Madrid, Luong N. Nguyen, Long D. Nghiem, Alex V. Chaves, and Peter J. Ralph. 2020. Management of enteric methanogenesis in ruminants by algal-derived feed additives. *Current Pollution Reports* 6 (3): 188–205. https://doi.org/10.1007/s40726-020-00151-7.

McDonald, Bruce A. 2014. Using dynamic diversity to achieve durable disease resistance in agricultural ecosystems. *Tropical Plant Pathology* 39 (3): 191–196. https://doi.org/10.1590/S1982-56762014000300001.

McDougall, Phillips. 2011. *The cost and time involved in the discovery, development and authorisation of a new plant biotechnology derived trait*. A consultancy study for Crop Life International, 1–24.

Mehrabi, Zia, Mollie J. McDowell, Vincent Ricciardi, Christian Levers, Juan Diego Martinez, Natascha Mehrabi, Hannah Wittman, Navin Ramankutty, and Andy Jarvis. 2020. The global divide in data-driven farming. *Nature Sustainability*: 1–7. https://doi.org/10.1038/s41893-020-00631-0.

Menchaca, A., P.C. dos Santos-Neto, A.P. Mulet, and M. Crispo. 2020. CRISPR in livestock: From editing to printing. *Theriogenology* 150: 247–254. https://doi.org/10.1016/j.theriogenology.2020.01.063.

Mihelcic, James R., Lauren M. Fry, and Ryan Shaw. 2011. Global potential of phosphorus recovery from human urine and feces. *Chemosphere*. https://doi.org/10.1016/j.chemosphere.2011.02.046.

Monteiro, Carlos A., J.-C. Moubarac, Geoffrey Cannon, Shu Wen Ng, and Barry Popkin. 2013. Ultra-processed products are becoming dominant in the global food system. *Obesity Reviews* 14 (Suppl. 2, November): 21–28. https://doi.org/10.1111/obr.12107.

Moraine, Marc, Michel Duru, Pip Nicholas, Philippe Leterme, and Olivier Therond. 2014. Farming system design for innovative crop-livestock integration in Europe. *Animal: An International Journal of Animal Bioscience* 8 (May): 1–14. https://doi.org/10.1017/S1751731114001189.

Moroz, Peter W., Oana Branzei, Simon C. Parker, and Edward N. Gamble. 2018. Imprinting with purpose: Prosocial opportunities and B Corp certification. *Journal of Business Venturing* 33 (2): 117–129. https://doi.org/10.1016/j.jbusvent.2018.01.003.

Mus, Florence, Matthew B. Crook, Kevin Garcia, Amaya Garcia Costas, Barney A. Geddes, Evangelia D. Kouri, Ponraj Paramasivan, et al. 2016. Symbiotic nitrogen fixation and the challenges to its extension to nonlegumes. *Applied and Environmental Microbiology* 82 (13): 3698–3710. https://doi.org/10.1128/AEM.01055-16.

Napier, Jonathan A., and Olga Sayanova. 2020. Nutritional enhancement in plants—Green and greener. *Current Opinion in Biotechnology* 61: 122–127. https://doi.org/10.1016/j.copbio.2019.12.010.

National Academies of Sciences, Engineering, and Medicine. 2016. *Genetically engineered crops: Experiences and prospects*. Washington, DC: National Academies Press.

NatureScot. 2020. Agroecological transitions—Case studies. Accessed November 7, 2020. https://www.nature.scot/agroecological-transitions-case-studies.

Nelson, Rebecca, Richard Coe, and Bettina I.G. Haussmann. 2019. Farmer research networks as a strategy for matching diverse options and contexts in smallholder agriculture. *Experimental Agriculture* 55 (S1): 125–144. https://doi.org/10.1017/S0014479716000454.

Nicholls, Robert J., Susan Hanson, Celine Herweijer, Nicola Patmore, Stéphane Hallegatte, Jan Corfee-Morlot, Jean Château, and Robert Muir-Wood. 2008. Ranking port cities with high exposure and vulnerability to climate extremes: Exposure estimates. OECD Environment Working Papers, No. 1, OECD Publishing, Paris. https://doi.org/10.1787/011766488208.

Oerke, Erich Christian. 2006. Crop losses to pests. *The Journal of Agricultural Science* 144 (1): 31–43. https://doi.org/10.1017/S0021859605005708.

Oladosu, Yusuff, Mohd Y. Rafii, Fatai Arolu, Samuel Chibuike Chukwu, Ismaila Muhammad, Isiaka Kareem, Monsuru Adekunle Salisu, and Ibrahim Wasiu Arolu. 2020. Submergence tolerance in rice: Review of mechanism, breeding and future prospects. *Sustainability* 12 (4). https://doi.org/10.3390/su12041632.

Organization for Economic Cooperation and Development (OECD). 2018. *States of fragility 2018*. Paris: OECD.

Organization for Economic Cooperation and Development (OECD). 2020. *Agricultural policy monitoring and evaluation 2020*. Paris: OECD Publishing. https://doi.org/10.1787/928181a8-en.

Orner, Kevin D., and James R. Mihelcic. 2018. A review of sanitation technologies to achieve multiple sustainable development goals that promote resource recovery. *Environmental Science: Water Research and Technology* 4 (1): 16–32. https://doi.org/10.1039/c7ew00195a.

Ortiz-Colón, Guillermo, Stephen J. Fain, Isabel K. Parés, Jaime Curbelo-Rodríguez, Esbal Jiménez-Cabán, Melvin Pagán-Morales, and William A. Gould. 2018. Assessing climate vulnerabilities and adaptive strategies for resilient beef and dairy 26 operations in the tropics. *Climatic Change* 146: 47–58. https://doi.org/10.1007/s10584-017-2110-1.

O'Sullivan, Aifric, Bethany Henrick, Bonnie Dixon, Daniela Barile, Angela Zivkovic, Jennifer Smilowitz, Danielle Lemay, J. William Martin, Bruce German, and Sara Elizabeth Schaefer. 2018. 21st century toolkit for optimizing population health through precision nutrition. *Critical Reviews in Food Science and Nutrition* 58 (17): 3004–3015. https://doi.org/10.1080/10408398.2017.1348335.

Pannico, A., C. El-Nakhel, M.C. Kyriacou, M. Giordano, S.R. Stazi, S. De Pascale, and Y. Rouphael. 2019. Combating micronutrient deficiency and enhancing food functional quality through selenium fortification of select lettuce genotypes grown in a closed soilless system. *Frontiers in Plant Science* 10 (November): 1–16. https://doi.org/10.3389/fpls.2019.01495.

Pardey, Philip G., Connie Chan-Kang, Steven P. Dehmer, and Jason M. Beddow. 2016. Agricultural R&D is on the move. *Nature News* 537 (7620): 301. https://doi.org/10.1038/537301a.

Parodi, A., A. Leip, I.J.M. De Boer, P.M. Slegers, F. Ziegler, E.H. Temme, M. Herrero, H. Tuomisto, H. Valin, C.E. Van Middelaar, and J.J.A. Van Loon. 2018. The potential of future foods for sustainable and healthy diets. *Nature Sustainability* 1 (12): 782–789.

Patel, Seema. 2019. Chapter 2—Insects as a Source of Sustainable Proteins. In *Proteins: Sustainable source, processing and applications*, ed. Charis M. Galanakis, 41–61. Academic Press. https://doi.org/10.1016/B978-0-12-816695-6.00002-7.

Pilet-Nayel, Marie-Laure, Benoît Moury, Valérie Caffier, Josselin Montarry, Marie-Claire Kerlan, Sylvain Fournet, Charles-Eric Durel, and Régine Delourme. 2017. Quantitative resistance to plant pathogens in pyramiding strategies for durable crop protection. *Frontiers in Plant Science* 8 (October). https://doi.org/10.3389/fpls.2017.01838.

Pinstrup-Andersen, Per. 2018. Is it time to take vertical indoor farming seriously? *Global Food Security* 17 (June): 233–235. https://doi.org/10.1016/j.gfs.2017.09.002.

Pixley, Kevin V., Jose B. Falck-Zepeda, Ken E. Giller, Leland L. Glenna, Fred Gould, Carol A. Mallory-Smith, David M. Stelly, and C. Neal Stewart Jr. 2019. Genome editing, gene drives, and synthetic biology: Will they contribute to disease-resistant crops, and who will benefit? *Annual Review of Phytopathology* 57: 165–188.

Place, Frank, Christopher B. Barrett, H. Ade Freeman, Joshua J. Ramisch, and Bernard Vanlauwe. 2003. Prospects for integrated soil fertility management using organic and inorganic inputs: Evidence from smallholder African agricultural systems. *Food Policy* 28 (4): 365–378.

Porciello, Jaron, Hale Ann Tufan, Jemimah Njuki, Paul Winters, Edward Mabaya, and Ronnie Coffman. 2020. Averting hunger in Sub-Saharan Africa requires data and synthesis. *Nature* 584 (7819): 37–40. https://doi.org/10.1038/d41586-020-02281-w.

Pretty, Jules, Tim G. Benton, Zareen Pervez Bharucha, Lynn V. Dicks, Cornelia Butler Flora, H. Charles, J. Godfray, Dave Goulson, et al. 2018. Global assessment of agricultural system redesign for sustainable intensification. *Nature Sustainability* 1 (8): 441. https://doi.org/10.1038/s41893-018-0114-0.

Pryce, Jennie E., and Mekonnen Haile-Mariam. 2020. Symposium review: Genomic selection for reducing environmental impact and adapting to climate change. *Journal of Dairy Science* 103 (6): 5366–5375. https://doi.org/10.3168/jds.2019-17732.

Qaim, Matin. 2016. *Genetically modified crops and agricultural development*. Palgrave Macmillan. https://doi.org/10.1057/9781137405722.

Raghunathan, Kalyani, Derek Headey, and Anna Herforth. 2020. Affordability of nutritious diets in rural India. *Food Policy* (2020): 101982. https://doi.org/10.1016/j.foodpol.2020.101982.

Reeves, T.G., and K. Cassaday. 2002. History and past achievements of plant breeding. *Australian Journal of Agricultural Research* 53 (8): 851–863. https://doi.org/10.1071/AR02038.

ReFEd. 2016. *A roadmap to reduce U.S. food waste by 20 percent*, 91. https://www.rockefellerfoundation.org/wp-content/uploads/ReFED_Roadmap_to_Reduce_Food_Waste.pdf.

Regis, Ed. 2019. *Golden rice: The imperiled birth of a GMO superfood*. Johns Hopkins University Press.

Reineke, Kai, and Alexander Mathys. 2020. Endospore inactivation by emerging technologies: A review of target structures and inactivation mechanisms. *Annual Review of Food Science and Technology* 11 (1): 255–274. https://doi.org/10.1146/annurev-food-032519-051632.

Reyes, Marcela, María Luisa Garmendia, Sonia Olivares, Claudio Aqueveque, Isabel Zacarías, and Camila Corvalán. 2019. Development of the Chilean front-of-package food warning label. *BMC Public Health* 19 (1): 906. https://doi.org/10.1186/s12889-019-7118-1.

Rosenblueth, Mónica, Ernesto Ormeño-Orrillo, Aline López-López, Marco A. Rogel, Blanca Jazmín Reyes-Hernández, Julio C. Martínez-Romero, Pallavolu M. Reddy, and Esperanza Martínez-Romero. 2018. Nitrogen fixation in cereals. *Frontiers in Microbiology* 9.https://doi.org/10.3389/fmicb.2018.01794.

Russel, Kory C., Kelvin Hughes, Mary Roach, David Auerbach, Andrew Foote, Sasha Kramer, and Raúl Briceño. 2019. Taking container-based sanitation to scale: Opportunities and challenges. *Frontiers in Environmental Science* 7 (November): 1–7. https://doi.org/10.3389/fenvs.2019.00190.

Routledge, Michael N., Sinead Watson, and Yun Yun Gong. 2016. Aflatoxin exposure and associated human health effects: A review of epidemiological studies. *Food Safety* 4 (1): 14–27. https://doi.org/10.14252/foodsafetyfscj.2015026.

Sainsbury, Emma, Roger Magnusson, Anne-Marie Thow, and Stephen Colagiuri. 2020. Explaining resistance to regulatory interventions to prevent obesity and improve nutrition: A case-study of a sugar-sweetened beverages tax in Australia. *Food Policy* 93 (May): 101904.

Salam, Rehana A., Jai K. Das, Wardah Ahmed, Omar Irfan, Sana Sadiq Sheikh, and Zulfiqar A. Bhutta. 2019. Effects of preventive nutrition interventions among adolescents on health and nutritional status in low- and middle-income countries: A systematic review and meta-analysis. *Nutrients* 12 (1). https://doi.org/10.3390/nu12010049.

Salzman, James, Genevieve Bennett, Nathaniel Carroll, Allie Goldstein, and Michael Jenkins. 2018. The global status and trends of payments for ecosystem services. *Nature Sustainability* 1 (3): 136–144. https://doi.org/10.1038/s41893-018-0033-0.

Sánchez-Bayo, Francisco, and Kris A. G. Wyckhuys. 2019. Worldwide decline of the entomofauna: A review of its drivers. *Biological Conservation* 232 (April): 8–27. https://doi.org/10.1016/j.biocon.2019.01.020.

Sato, Toshio, Manzoor Qadir, Sadahiro Yamamoto, Tsuneyoshi Endo, and Zahoor Ahmad. 2013. Global, regional, and country level need for data on wastewater generation, treatment, and use. *Agricultural Water Management* 130 (December): 1–13. https://doi.org/10.1016/j.agwat.2013.08.007.

Schlesinger, William H., and Ronald Amundson. 2019. Managing for soil carbon sequestration: Let's get realistic. *Global Change Biology* 25 (2): 386–389.

Schumpeter, Joseph A. 1942. *Capitalism, socialism, and democracy*. New York: Harper and Brothers.

Schut, Antonius G.T., and Ken Giller. 2020. Soil-based, field-specific fertilizer recommendations are a pipe-dream. *Geoderma* 380: 114680.

Scott, C., B. Hawkins, and C. Knai. 2017. Food and beverage product reformulation as a corporate political strategy. *Social Science & Medicine* 172: 37–45. https://doi.org/10.1016/j.socscimed.2016.11.020.

Searchinger, Timothy D., Chris Malins, Patrice Dumas, David Baldock, Joe Glauber, Thomas Jayne, Jikun Huang, and Paswell Marenya. 2020. *Revising public agricultural support to mitigate climate change*. Washington, DC: World Bank.

Seferidi, Paraskevi, Gyorgy Scrinis, Inge Huybrechts, Jeremy Woods, Paolo Vineis, and Christopher Millett. 2020. The neglected environmental impacts of ultra-processed foods. *The Lancet Planetary Health* 4 (10): e437–e438.

Sen, Amartya. 1981. *Poverty and famines: An essay on entitlement and deprivation*. Oxford: Oxford University Press.

Sergelidis, Daniel. 2019. Lab grown meat: The future sustainable alternative to meat or a novel functional food? *Biomedical Journal of Scientific & Technical Research* 17 (1): 12440–12444. https://doi.org/10.26717/BJSTR.2019.17.002930.

Shah, Mahfuzur Rahman, Giovanni Antonio Lutzu, Asraful Alam, M.A. Pallab Sarker, Kabir Chowdhury, Ali Parsaeimehr, Yuanmei Liang, and Maurycy Daroch. 2018. Microalgae in aquafeeds for a sustainable aquaculture industry. *Journal of Applied Phycology* 30 (1): 197–213. https://doi.org/10.1007/s10811-017-1234-z.

Sheahan, Megan, Christopher B. Barrett, and Casey Goldvale. 2017. Human health and pesticide use in Sub-Saharan Africa. *Agricultural Economics* 48 (S1): 27–41. https://doi.org/10.1111/agec.12384.

Shen, Jianbo, Chunjian Li, Guohua Mi, Long Li, Lixing Yuan, Rongfeng Jiang, and Fusuo Zhang. 2013. Maximizing root/rhizosphere efficiency to improve crop productivity and nutrient use efficiency in intensive agriculture of China. *Journal of Experimental Botany* 64: 1181–1192. https://doi.org/10.1093/jxb/ers342.

Shen, Jianbo, Zhenling Cui, Yuxin Miao, Guohua Mi, Hongyan Zhang, Mingsheng Fan, Chaochun Zhang, et al. 2013. Transforming agriculture in China: From solely high yield to both high yield and high resource use efficiency. *Global Food Security* 2 (1): 1–8. https://doi.org/10.1016/j.gfs.2012.12.004.

Shepon, Alon, Jessica A. Gephart, Patrik John Gustav Henriksson, Robert Jones, Khondker Murshed-e-Jahan, Gidon Eshel, and Christopher D. Golden. 2020. Reorientation of aquaculture production systems can reduce environmental impacts and improve nutrition security in Bangladesh. *Nature Food* 1 (10): 640–647. https://doi.org/10.1038/s43016-020-00156-x.

Shieber, Jonathan. 2020. Alternative protein companies have raised a whopping $1.5 billion through July of this year. *TechCrunch* (blog), September 17. https://social.techcrunch.com/2020/09/17/alternative-protein-companies-have-raised-a-whopping-1-5-billion-through-july-of-this-year/.

Sibhatu, Kibrom T., Vijesh V. Krishna, and Matin Qaim. 2015. Production diversity and dietary diversity in smallholder farm households. *Proceedings of the National Academy of Sciences* 112 (34): 10657–10662.

Sibhatu, Kibrom T., and Matin Qaim. 2018. Meta-analysis of the association between production diversity, diets, and nutrition in smallholder farm households. *Food Policy* 77 (May): 1–18. https://doi.org/10.1016/j.foodpol.2018.04.013.

Siegrist, Michael, and Christina Hartmann. 2020. Impact of sustainability perception on consumption of organic meat and meat substitutes. *Appetite* 132 (January): 196–202. https://doi.org/10.1016/j.appet.2018.09.016.

Smetana, Sergiy, Michael Sandmann, Sascha Rohn, Daniel Pleissner, and Volker Heinz. 2017. Autotrophic and heterotrophic microalgae and cyanobacteria cultivation for food and feed: Life cycle assessment. *Bioresource Technology* 245 (December): 162–170. https://doi.org/10.1016/j.biortech.2017.08.113.

Springmann, Marco, Michael A. Clark, Mike Rayner, Peter Scarborough, and Patrick Webb. 2021. The global and regional costs of healthy and sustainable dietary patterns: A modelling study. *The Lancet Planetary Health* 5 (11): e797–e807. https://doi.org/10.1016/S2542-5196(21)00251-5.

Smetana, Sergiy, Eric Schmitt, and Alexander Mathys. 2019. Sustainable use of Hermetia illucens insect biomass for feed and food: Attributional and consequential life cycle assessment. *Resources, Conservation and Recycling* 144 (February): 285–296. https://doi.org/10.1016/j.resconrec.2019.01.042.

Smyth, Stuart J. 2020. Regulatory barriers to improving global food security. *Global Food Security* 26 (September): 100440. https://doi.org/10.1016/j.gfs.2020.100440.

Sola, P., B.M. Mvumi, J.O. Ogendo, et al. 2014. Botanical pesticide production, trade and regulatory mechanisms in sub-Saharan Africa: Making a case for plant-based pesticidal products. *Food Security* 6: 369–384. https://doi.org.proxy.library.cornell.edu/10.1007/s12571-014-0343-7.

Stenek, Vladimir, Jean-Christophe Amado, Richenda Connell, Olivia Palin, Stewart Wright, Ben Pope, John Hunter, James McGregor, Will Morgan, Ben Stanley, Richard Washington, Diana Liverman, Hope Sherwin, Paul Kapelus, Carlos Andrade, and José D. Pabon. 2011. Climate risk and business: Ports; Terminal Maritimo Muelles el Bosque, Cartagena, Colombia, International Finance Corporation, Washington.

Stevenson, Philip C., Steven R. Belmain, and Murray B. Isman. 2020. Pesticidal plants: From smallholder use to commercialization. Basel: MDPI.

Stiglitz, Joseph E., Amartya Sen, and Jean-Paul Fitoussi. 2010. *Mismeasuring our lives: Why GDP doesn't add up*. New York: New Press.

Strandén, Ismo, Juha Kantanen, Isa-Rita M. Russo, Pablo Orozco-terWengel, Michael W. Bruford, and the Climgen Consortium. 2019. Genomic selection strategies for breeding adaptation and production in dairy cattle under climate change. *Heredity* 123 (3): 307–317. https://doi.org/10.1038/s41437-019-0207-1.

Stokstad, Erik. 2019. After 20 years, golden rice nears approval. *Science* 366 (6468): 934–934. https://doi.org/10.1126/science.366.6468.934.

Sunstein, Cass R. Are food labels good? *Food Policy* (in print): 101984.

Taillie, Lindsey Smith, Marcela Reyes, M. Arantxa Colchero, Barry Popkin, and Camila Corvalán. 2020. An evaluation of Chile's law of food labeling and advertising on sugar-sweetened beverage purchases from 2015 to 2017: A before-and-after study. *PLoS Medicine* 17 (2): e1003015.

Talakayala, Ashwini, Sumalatha Katta, and Mallikarjuna Garladinne. 2020. Genetic engineering of crops for insect resistance: An overview. *Journal of Biosciences* 45 (1). https://doi.org/10.1007/s12038-020-00081-y.

Thow, Anne Marie, Stephen Jan, Stephen Leeder, and Boyd Swinburn. 2010. The effect of fiscal policy on diet, obesity and chronic disease: A systematic review. *Bulletin of the World Health Organization* 88: 609–614.

Thow, Anne Marie, Shauna Downs, and Stephen Jan. 2014. A systematic review of the effectiveness of food taxes and subsidies to improve diets: Understanding the recent evidence. *Nutrition reviews* 72 (9): 551–565. https://doi.org/10.1111/nure.12123.

Tobin, D., K. Jones, and B.C. Thiede. 2019. Does crop diversity at the village level influence child nutrition security? Evidence from 11 Sub-Saharan African countries. *Population and Environment.* 41 (2): 74–97. https://doi.org/10.1007/s11111-019-00327-4.

Toro-Martín, Juan de, Benoit J. Arsenault, Jean-Pierre Després, and Marie-Claude Vohl. 2017. Precision nutrition: A review of personalized nutritional approaches for the prevention and management of metabolic syndrome. *Nutrients* 9 (8). https://doi.org/10.3390/nu9080913.

Tubb, Catherine, and Tony Seba. 2019. *Rethinking food and agriculture 2020–2030: The second domestication of plants and animals, the disruption of the cow, and the collapse of industrial livestock farming*. San Francisco: Rethinkx.

United Nations Environmental Program. (2013). Guidelines for national waste management strategies moving from challenges to opportunities. ISBN 978-92-807-3333-4.

Vallée Marcotte, Bastien, Frédéric Guénard, Simone Lemieux, Patrick Couture, Iwona Rudkowska, Philip C. Calder, Anne Marie Minihane, and Marie-Claude Vohl. 2019. Fine mapping of genome-wide association study signals to identify genetic markers of the plasma triglyceride response to an omega-3 fatty acid supplementation. *The American Journal of Clinical Nutrition* 109 (1): 176–185. https://doi.org/10.1093/ajcn/nqy298.

Van Deynze, Allen, Pablo Zamora, Pierre-Marc Delaux, Cristobal Heitmann, Dhileepkumar Jayaraman, Shanmugam Rajasekar, Danielle Graham, et al. 2018. Nitrogen fixation in a landrace of maize is supported by a mucilage-associated diazotrophic microbiota. *PLoS Biology* 16 (8): e2006352. https://doi.org/10.1371/journal.pbio.2006352.

Van Eck, J. 2020. Applying gene editing to tailor precise genetic modifications in plants. *The Journal of Biological Chemistry*. https://doi.org/10.1074/jbc.REV120.010850.

Van Esse, H. Peter, T. Lynne Reuber, and Dieuwertje van der Does. 2020. Genetic modification to improve disease resistance in crops. *The New Phytologist* 225 (1): 70–86. https://doi.org/10.1111/nph.15967.

Van Etten, Jacob, Kaue de Sousa, Amilcar Aguilar, Mirna Barrios, Allan Coto, Matteo Dell'Acqua, Carlo Fadda, et al. 2019. Crop variety management for climate adaptation supported by citizen science. *Proceedings of the National Academy of Sciences of the United States of America* 10: 4194. https://doi.org/10.1073/pnas.1813720116.

Van Grinsven, Hans J. M., Mike Holland, Brian H. Jacobsen, Zbigniew Klimont, Mark A. Sutton, and W. Jaap Willems. 2013. Costs and benefits of nitrogen for Europe and implications for mitigation. *Environmental Science & Technology* 47 (8): 3571–79. https://doi.org/10.1021/es303804g.

van Huis, Arnold. 2013. Potential of insects as food and feed in assuring food security. *Annual Review of Entomology* 58 (1): 563–583. https://doi.org/10.1146/annurev-ento-120811-153704.

van Lenteren, Joop C., Karel Bolckmans, Jürgen Köhl, Willem J. Ravensberg, and Alberto Urbaneja. 2018. Biological control using invertebrates and microorganisms: Plenty of new opportunities. *BioControl* 63 (1): 39–59. https://doi.org/10.1007/s10526-017-9801-4.

Vandermeer, John. 1997. Syndromes of production: An emergent property of simple agroecosystem dynamics. *Journal of Environmental Management* 51: 59–72.

Vandevijvere, Stefanie, and Lana Vanderlee. 2019. Effect of formulation, labelling, and taxation policies on the nutritional quality of the food supply. *Current Nutrition Reports* 8 (3): 240–249. https://doi.org/10.1007/s13668-019-00289-x.

Vandevijvere, Stefanie, Lindsay M. Jaacks, Carlos A. Monteiro, Jean-Claude Moubarac, Martin Girling-Butcher, Arier C. Lee, An Pan, James Bentham, and Boyd Swinburn. 2019. Global trends in ultraprocessed food and drink product sales and their association with adult body mass index trajectories. *Obesity Reviews*.

Vanlauwe, B., A. Bationo, J. Chianu, K.E. Giller, R. Merckx, U. Mokwunye, O. Ohiokpehai, P. Pypers, R. Tabo, K.D. Shepherd, and E.M.A. Smaling. 2010. Integrated soil fertility management: Operational definition and consequences for implementation and dissemination. *Outlook on Agriculture* 39 (1): 17–24.

Vanlauwe, B., K. Descheemaeker, K.E. Giller, J. Huising, R. Merckx, G. Nziguheba, J. Wendt, and S. Zingore. 2015. Integrated soil fertility management in sub-Saharan Africa: Unravelling local adaptation. *The Soil* 1 (1): 491–508.

Vicente, Emilio Jimenez, and Dennis R. Dean. 2017. Keeping the nitrogen-fixation dream alive. *Proceedings of the National Academy of Sciences of the United States of America* 114 (12): 3009–3011.

Wagner, Erika B. 2011. Why prize? The surprising resurgence of prizes to stimulate innovation. *Research-Technology Management* 54 (6): 32–36. https://doi.org/10.5437/08956308X5406013.

Wang, Liyang, and Jianbo Shen. 2019. Root/rhizosphere management for improving phosphorus use efficiency and crop productivity. *Better Crops* 103 (1): 36–39. https://doi.org/10.24047/BC103136.

Wang, Xin, William R. Whalley, Anthony J. Miller, Philip J. White, Fusuo Zhang, and Jianbo Shen. 2020. Sustainable cropping requires adaptation to a heterogeneous rhizosphere. *Trends in Plant Science*. https://doi.org/10.1016/j.tplants.2020.07.006.

Wen, Zhihui, Hongbo Li, Qi Shen, Xiaomei Tang, Chuanyong Xiong, Haigang Li, Jiayin Pang, Megan H. Ryan, Hans Lambers, and Jianbo Shen. 2019. Tradeoffs among root morphology, exudation and mycorrhizal symbioses for phosphorus-acquisition strategies of 16 crop species. *New Phytologist* 223: 882–895. https://doi.org/10.1111/nph.15833.

Westerberg, Arthur W. 2004. A retrospective on design and process synthesis. *Computers & Chemical Engineering* 28 (4): 447–458. https://doi.org/10.1016/j.compchemeng.2003.09.029.

Wezel, A., and V. Soldat. 2009. A quantitative and qualitative historical analysis of the scientific discipline of agroecology. *International Journal of Agricultural Sustainability* 7 (1): 3–18. https://doi.org/10.3763/ijas.2009.0400.

Wezel, Alexander, Barbara Gemmill Herren, Rachel Bezner Kerr, Edmundo Barrios, André Luiz Rodrigues Gonçalves, and Fergus Sinclair. 2020. Agroecological principles and elements and their implications for transitioning to sustainable food systems: A review. *Agronomy for Sustainable Development*. https://doi.org/10.1007/s13593-020-00646-z.

WHO and UNICEF. 2017. *Progress on drinking water, sanitation and hygiene: 2017 update and SDG baseline*. Switzerland: World Health Organization (WHO) and United Nations Children's Fund (UNICEF).

Willett, Walter, Johan Rockström, Brent Loken, Marco Springmann, Tim Lang, Sonja Vermeulen, Tara Garnett, et al. 2019. Food in the Anthropocene: The EAT–Lancet Commission on healthy diets from sustainable food systems. *The Lancet* 393 (10170): 447–492. https://doi.org/10.1016/S0140-6736(18)31788-4.

Wing, Rod A., Michael D. Purugganan, and Qifa Zhang. 2018. The rice genome revolution: From an ancient grain to green super rice. *Nature Reviews Genetics* 19 (8): 505–517. https://doi.org/10.1038/s41576-018-0024-z.

World Bank. 2020. *State and trends of carbon pricing 2020*. Washington, DC: World Bank.
WWF. 2020. *Indoor soilless farming: Phase I: Examining the industry and impacts of controlled environment agriculture*. Washington, DC: World Wildlife Fund (WWF).
Yang, Fan, Min Zhang, and Bhesh Bhandari. 2017. Recent development in 3D food printing. *Critical Reviews in Food Science and Nutrition* 57 (14): 3145–3153. https://doi.org/10.1080/10408398.2015.1094732.
Yeh, Christopher, Anthony Perez, Anne Driscoll, George Azzari, Zhongyi Tang, David Lobell, Stefano Ermon, and Marshall Burke. 2020. Using publicly available satellite imagery and deep learning to understand economic well-being in Africa. *Nature Communications* 11 (1): 2583. https://doi.org/10.1038/s41467-020-16185-w.
Yuan, Lixia, and Runzhi Li. 2020. Metabolic engineering a model oilseed *Camelina sativa* for the sustainable production of high-value designed oils. *Frontiers in Plant Science* 11 (February): 1–14. https://doi.org/10.3389/fpls.2020.00011.
Zaidi, Syed Shan-e-Ali., Hervé Vanderschuren, Matin Qaim, Magdy M. Mahfouz, Ajay Kohli, Shahid Mansoor, and Mark Tester. 2019. New plant breeding technologies for food security. *Science* 363 (6434): 1390–1391. https://doi.org/10.1126/science.aav6316.
Zhang, Deshan, Chaochun Zhang, Xiaoyan Tang, Haigang Li, Fusuo Zhang, Zed Rengel, William R. Whalley, William J. Davies, and Jianbo Shen. 2016. Increased soil P availability induced by faba bean root exudation stimulates root growth and P uptake in neighbouring maize. *New Phytologist* 209: 823–883. https://doi.org/10.1111/nph.13613.
Zhang, W.F., G.X. Cao, X.L. Li, H.Y. Zhang, C. Wang, Q.C. Liu, X.P. Chen, Z.L. Cui, J.B. Shen, R.F. Jiang, G.H. Mi, Y.X. Miao, F.S. Zhang, and Z.X. Dou. 2016. Closing yield gaps in China by empowering smallholder farmers. *Nature* 537 (7622): 671–674. https://doi.org/10.1038/nature19368.
Zhang, Deshan, Yang Lyu, Hongbo Li, Xiaoyan Tang, Ran Hu, Zed Rengel, Fusuo Zhang, et al. 2020. Neighbouring plants modify maize root foraging for phosphorus: Coupling nutrients and neighbours for improved nutrient-use efficiency. *New Phytologist* 226: 244–253. https://doi.org/10.1111/nph.16206.
Zulkifli, Masagos. 2020. Speech at Launch for 2020: Singapore food story, February 10. Accessed November 6, 2020. http://sglinks.news/mewr/pr/speech-mr-masagos-zulkifli-minister-environment-water-resources-1e6b08e.

Socio-Technical Innovation Bundles Tailored to Distinct Agri-Food Systems

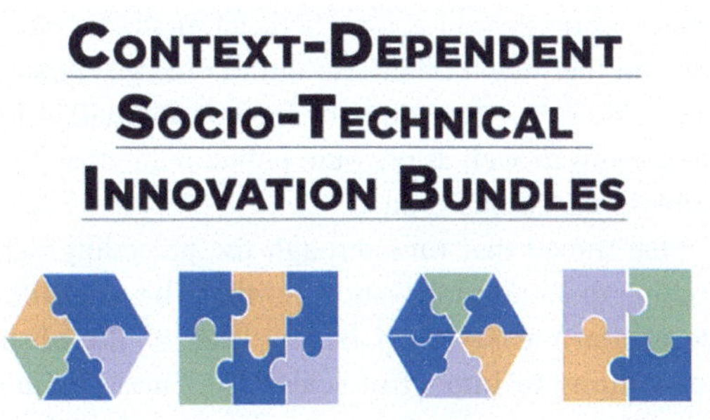

Scientific discovery is neither linear nor predictable. The time it takes to develop breakthrough technologies varies enormously among application domains. Some basic scientific discoveries remain elusive and will need continued, concerted funding and attention in the years and decades ahead. In some cases, the stumbling block is the scientific advancement per se, when important discoveries along the path towards technological readiness have not yet been made. This has been the case, for example, with numerous vaccines, both for humans (e.g., malaria, HIV) and for livestock (e.g., East Coast fever, trypanosomiasis, African swine fever). Research teams must sometimes work for several decades on the science necessary for a breakthrough discovery that can lead to a demonstrably effective, scalable product or impact. Similarly, several emerging options that could revolutionize crop yields (e.g., reconfiguring photosynthetic pathways for greater efficiency, nitrogenase in cereals) have remained elusive but continue to show sufficient promise to merit generous R&D investment. But even when breakthroughs occur, the time to market may be long, often decades.

Promising innovations often do not gain traction, not because the underlying science has proved too difficult but, rather, because the enabling environment

C. B. Barrett et al., *Socio-Technical Innovation Bundles for Agri-Food Systems Transformation*, Sustainable Development Goals Series,
https://doi.org/10.1007/978-3-030-88802-2_7

essential to development and diffusion is lacking. Most breakthrough science requires financial, institutional, and sociopolitical support in order to advance through pilot stages to achieve impact at scale. It is therefore essential to identify the socio-technical bundles that combine social and scientific elements to unlock the transformative potential of emergent technologies.

Indeed, throughout history **all dramatic new technological inventions and impactful innovations have been combinatorial**, brought about through the intentional combination of different prior discoveries with the express intent of solving a human need (Arthur 2009). Transformative innovation therefore necessarily involves bundles of (1) scientific and engineering advances that improve the attributes of goods and processes; (2) public policies that induce appropriate behaviors by private actors, both internalizing externalities and advancing coordination that might otherwise fail to emerge spontaneously; and (3) informal private behaviors—the culture of food, if you will—that incentivize and help diffuse innovations as well as pressure public policymakers. Transformation thus requires multiple transitions at once.

One thread that runs through the preceding, lengthy discussion of scores of exciting emergent innovations is **that the scientific challenges, while formidable in many cases, may be the least of the obstacles to bringing promising innovations to impactful scale**. The "best" or most scientifically elegant technologies only occasionally prevail, often floundering due to cultural, economic, ethical, or political counter-pressures. The agri-food transformations that capture attention are often too narrowly associated with a particular emblematic technology that was central to their success. **The sociocultural, policy, and/or institutional changes that enable that new science to turn into transformative technologies are commonly overlooked but are equally important. Hence the importance of bundling**.

INDEED, THROUGHOUT HISTORY ALL DRAMATIC NEW TECHNOLOGICAL INVENTIONS AND IMPACTFUL INNOVATIONS HAVE BEEN COMBINATORIAL, BROUGHT ABOUT THROUGH THE INTENTIONAL COMBINATION OF DIFFERENT PRIOR DISCOVERIES WITH THE EXPRESS INTENT OF SOLVING A HUMAN NEED.

For example, the Asian Green Revolution, which genuinely transformed Asia's AFSs, was not just a result of the development of input-responsive high-yielding crop varieties, although these are the emblematic technology of the era. The transformation required a whole ecosystem of structures and institutions to make it work, and this took considerable time to emerge and develop, at least a decade. In the case of the Asian Green Revolution, the ecosystem included public investment in irrigation, transportation and communications infrastructure, input supply arrangements, public pricing, and procurement systems; a set of shared values among a group of philanthropic agencies, government bureaucrats, and international and local scientists to both develop and promote the new technology; and commitments to making the technology an international public good freely available to breeding programs worldwide. Nearly half a century later, these same technologies have failed to transform the AFSs of sub-Saharan Africa precisely because this wider enabling environment has yet to emerge.

Other examples reinforce this point. For example, the 2011 declaration of the eradication of rinderpest (cattle plague)—an animal disease with enormous adverse impact over centuries, especially in sub-Saharan Africa—featured a new vaccine as an emblematic technology but relied equally on a complex ecosystem of global scientific cooperation, cold chain distribution infrastructure, national policy and regulatory changes, awareness campaigns, and internationally coordinated vaccination programs. Like the Green Revolution, it also depended on generous, non-commercial financing and unencumbered intellectual property rights on the vaccine.

As was clear in our earlier example of the simple comparison between rice genetics discoveries—the IR8, IR36, and IR64 varieties of the Green Revolution versus contemporary golden rice—"novel technologies alone are not sufficient to drive agri-food system transformations; instead, they must be accompanied by a wide range of social and institutional factors that enable their deployment" (Herrero et al. 2020, p. 267). Despite having viable transgenic rice varieties containing high levels of beta carotene for more than a decade, these varieties are yet to be produced by farmers independent of scientific trials, let alone consumed by the vitamin A–deficient populations for whom they were developed. A critical missing part of the ecosystem was social license, with major political and ethical opposition emerging in several target countries (Regis 2019).

These successes and failures led Herrero et al. (2020) to describe eight essential elements for accelerating systematic transformation in AFSs (left panel of Fig. 1). These actions complete the socio-cultural fabric of the enabling environment for increasing the chances that promising technologies get adapted to fit a given context, adopted by many, and ultimately scale to achieve the desired societal impacts. Which elements most impactfully combine with which technology

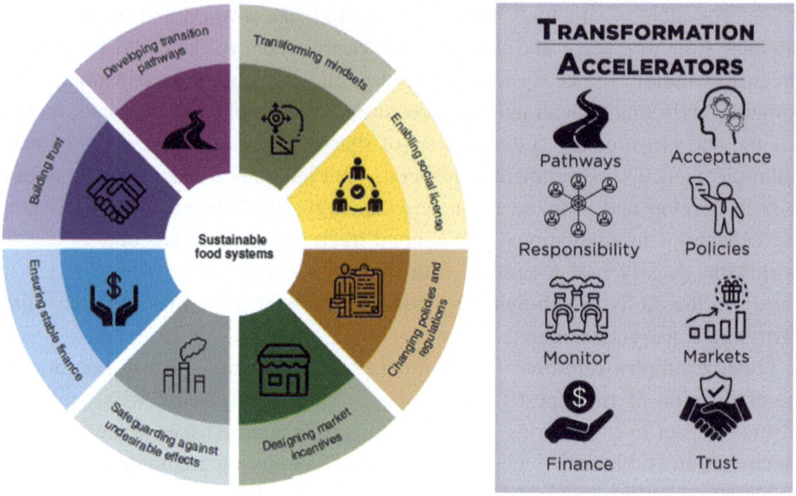

Fig. 1 Essential elements for accelerating the systemic transformation of agri-food systems (Lefthand figure reproduced from Herrero et al. 2020)

depends fundamentally on the context and the technology. But those combinations do not occur without human agency. The eight "transformation accelerators" depicted in Fig. 1 and Herrero et al. (2020) are all human actions: building trust, transforming mindsets, designing market incentives, etc.

We therefore emphasize socio-technical innovation bundles as appropriately contextualized combinations of science and technology advances that, when combined with specific, appropriate institutional or policy adaptations, exhibit particular promise for advancing one or more design objectives in a particular setting. **The task of discovering, adapting, and scaling beneficial innovations is as much one for humanists and social scientists as it is for engineers and natural scientists**. Agents throughout AVCs play an active role. Innovation is not just the business of engineers and scientists who think of R&D as their bread-and-butter activities. Table 1 works out a stylized example of the articulation of the need for these accelerators for two promising new upstream technologies described earlier: nitrogen-fixing cereals and circular (livestock) feeds.[1]

[1] The specifics of these cases are described in detail in Herrero et al. (2021).

Table 1 Accelerators for two promising agri-food technologies

Elements for AFS Transformation	INNOVATIONS	
	Circular Feeds (Microbial Protein from Organic Waste Streams)	Nitrogen Fixation in Cereals
Building Trust amongst Actors in the Food System  Vision and Values	**True for all Case Studies** • Trust building of "profit with a purpose" or "system positive benefits" • Transparent production, distribution, and management processes • Trust in regulatory enforcement of environmental, health, and safety standards	
	Specific to Case Study Technology	
	• Developing bio-safe production processes that ensure products are clean and safe to use/consume throughout the value chain (e.g. animals, operators, and consumers)	• Developing and confirming reliable nitrogen fixation and protein content in cereals • Ensuring the products are cost-effective for farmers and of high food safety and quality for consumers
Transforming Mindsets  Acceptance	**True for all Case Studies** • Acceptance of highly technological production and handling of food and feeds • Investment in education to increase awareness and appropriate use of new tech	
	Specific to Case Study Technology	
	• New by-product paradigm: waste of all types becomes input to other processes • Acceptance that feed can be produced from a range of organic waste streams, including animal and human waste	• Increased acceptance of applications of genetic modification/ gene transfer • Adjusted agricultural management practices to account for new bio-chemical requirements of advanced crops

(continued)

Table 1 (continued)

Elements for AFS Transformation	INNOVATIONS	
	Circular Feeds (Microbial Protein from Organic Waste Streams)	Nitrogen Fixation in Cereals
ENABLING SOCIAL LICENSE/STAKEHOLDER DIALOGUE  Responsibility	**True for all Case Studies** • Engage with stakeholders, including consumers, workers, and producers, to ensure technologies are developed and implemented transparently	
	Specific to Case Study Technology	
	• Deepen collaboration and cooperation between agriculture, and sanitation and waste management sectors to better understand each other's needs and social obligations	• Ensure quality of new crops as good as, or better than, alternatives • Demonstrate improved environmental profile that reduces input use/waste • Avoid vertical integration models that overly concentrate market power
ENSURING STABLE FINANCE  Explore and Pilot	**True for all Case Studies** • Clear and stable medium- to long-term goals adopted to signal to stakeholders the direction of change to reorient investment portfolios • Government soft loans, guarantees, and tax breaks linked to SDG/ESG performance • ESG public and private financing encouraged • New infrastructure investments based on long-term financing carried out • Given that early adopters are typically better off, financing that does not reinforce existing inequalities • Alternative funding mechanisms (e.g., AMCs, prizes) piloted to promote innovations that advance social and environmental objectives	
	Specific to Case Study Technology	
	• Prioritize funding to develop waste processing in diverse locations • Coordinate investments in sanitation and hygiene compatible w/ emerging waste processing technologies	• Promote open-access IP to increase access to novel crops for varied applications and business models

(continued)

Table 1 (continued)

Elements for AFS Transformation	INNOVATIONS	
	Circular Feeds (Microbial Protein from Organic Waste Streams)	Nitrogen Fixation in Cereals
Designing Market Incentives  Spread Cost and Risk	**True for all Case Studies** • Targeted fiscal and trade policies to ensure a viable, scalable initial market • Improved cost of externalities (environmental, social, health, etc.) at source to facilitate the competitiveness of new approaches	
	Specific to Case Study Technology	
	• Increase the cost of waste to encourage alternative use (e.g. increase waste handling fees) • Provide price supports for key inputs to reduce production costs • Target support to conventional feed sectors to transition to alternative production and land use	• Tax nitrogen leaching per the polluter pays principle to encourage uptake • Incentivize seed distribution networks to promote equitable farmer access • Develop mechanisms to repurpose N-fertilizer capital towards other economically and socially viable uses
Changing Policies and Regulations  Expectations Of Support	**True for all Case Studies** • Revised policies ensure effective oversight of new technologies and industries • Streamlined/coherent environmental, health, and safety regulations enacted throughout AFS • Policies targeted at reducing economic and bureaucratic constraints to technological adoption and diffusion	
	Specific to Case Study Technology	
	• Create circular feed targets for domestic animal diets • Improve coordination of waste processing and transport • Waste and agriculture authorities coordinate by-product disposal	• Optimize IP rights to facilitate diffusion of new technologies • Co-develop input supply markets with private industry

(continued)

Table 1 (continued)

Elements for AFS Transformation	INNOVATIONS	
	Circular Feeds (Microbial Protein from Organic Waste Streams)	Nitrogen Fixation in Cereals
Safeguarding Against Undesirable Effects  Monitor and Correct	**True for all Case Studies** • Capable, independent regulatory bodies transparently enforcing standards/rules • Inter-governmental agreements on environmental and labor standards for technology transfer and trade • Requirements for impact assessment, free prior informed consent, and other safeguarding principles for foreign direct investment • Enhanced mandatory ESG disclosure and SDG/SBT reporting • Increased ESG screening/reporting by financial institutions	
	Specific to Case Study Technology	
	• Identify potential zoonoses and chemical contamination sources • Disincentivize excess waste output • Monitor for downstream environmental and social impacts (e.g., increased production and consumption of livestock products)	• Monitor land use to ensure improving environmental footprint of the AFS • Monitor adverse impacts on biodiversity (pollinators, etc.) and agro-biodiversity (local varieties) to boost adoption of novel crops • Monitor soil nitrogen levels and tax surplus nitrogen to avoid over-fixation
Developing Transition Pathways  How and When	**True for all Case Studies** • Integrate the previous elements into an integrated implementation plan • Design transition pathways that not only promote winners, but ensure that those disadvantaged by change can also benefit from the fruits of innovation • Recognize there are no perfect solutions (let not perfection be the enemy of the good); prepare to course correct as unexpected consequences are identified • Focus not on specific technologies but on achieving AFS outcomes • Make local, national, and international commitment with appropriate resource allocation	

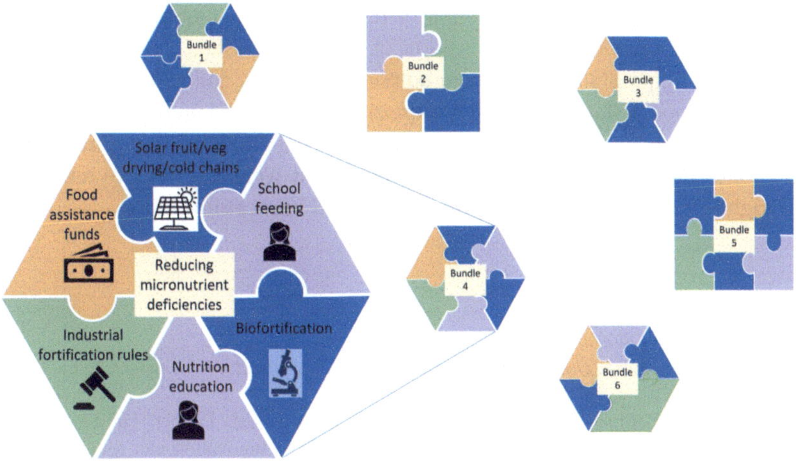

Fig. 2 Socio-technical bundles fit for purpose to an objective and context

Even with appropriately contextualized use of accelerators to enhance uptake of a given technology, many objectives require multiple, complementary interventions and the environment to support those multiple interventions. These often originate in different scientific spheres. A distinct set of multiple, mutually reinforcing innovations may be needed to achieve meaningful results at scale for a given design objective in a particular context. This, too, implies a need for contextualized socio-technical bundling of innovations, albeit for a slightly different purpose than for fostering and accelerating uptake of a given technology.

Figure 2 illustrates the case. Puzzle pieces represent innovations, which draw on different (natural or social) science-based methods (represented by different colors) to generate products, processes, or policies with distinct designs and purposes (represented by different shapes). These combine into different composite shapes to fit the people, place, and time. In this stylized figure, six distinct bundles are developed for half a dozen different objectives and AFS application domains. The right combination for one specific objective—in the enlarged case of bundle 4, reducing micronutrient (i.e., mineral and vitamin) deficiencies in a remote rural and traditional AFS—will differ from the bundle needed in other cases. Progress may require some combination of scientific advances (e.g., genetic improvement of crops through biofortification or inexpensive off-grid solar-powered fruit and vegetable drying and refrigeration technologies), financing (e.g., food assistance funding to enable poor consumers to afford a more diverse, nutrient-rich diet), legislation or regulation (e.g., required iodization of manufactured salt or folate fortification of flour and pasta), and policies (e.g., school feeding programs that feature nutrient-rich foods, and nutrition education to promote food culture, dietary diversity, and healthful food preparation and storage). The key point is that **science and engineering can design and adapt the raw materials, but ultimately stakeholders must work together to assemble the right bits into fit-for-purpose combinatorial innovations**.

References

Arthur, W. Brian. 2009. *The nature of technology: What it is and how it evolves*. New York: Free Press.

Herrero, Mario, Philip K. Thornton, Daniel Mason-D'Croz, Jeda Palmer, Tim G. Benton, Benjamin L. Bodirsky, Jessica R. Bogard, et al. 2020. Innovation can accelerate the transition towards a sustainable food system. *Nature Food* 1 (5): 266–272. https://doi.org/10.1038/s43016-020-0074-1.

Herrero, Mario, Philip K. Thornton, Daniel Mason-D'Croz, J. Palmer, B. L. Bodirsky, P. Pradhan, C. B. Barrett, T. G. Benton, et al. 2021. Articulating the impact of food systems innovation on the Sustainable Development Goals. *The Lancet Planetary Health* 5 (1): e50–e62. https://doi.org/10.1016/S2542-5196(20)30277-1.

Regis, Ed. 2019. *Golden rice: The imperiled birth of a GMO superfood*. Johns Hopkins University Press.

Impact Pathways

The complex pathways from innovation to impact mean that unintended spillover effects on non-target objectives are always likely. This generates a third reason—in addition to accelerators and complementarity in pursuit of target objectives—why socio-technical bundles are important. Herrero et al. (2020, 2021) demonstrated that food systems innovations can have mostly neutral or positive effects on the food systems SDGs (left-hand panel of Fig. 1). The likely impacts on non-food system SDGs, however, are more variable and not always positive (right-hand panel of Fig. 1). This was especially true for the SDGs concerned with growth (SDG 8); equity (SDG 10); and peace, justice, and strong institutions (SDG 16). Particularly those technologies related to digital agriculture, access to inputs, or increases in resource use efficiency could lead to significant winners and losers both, where the capacity of implementing these technologies—which require more education, good access to finance, and systems geared towards commercialization—might only allow some actors to engage beneficially.

These diverse spillover effects obviously depend on the specific type of technological intervention. As Herrero et al. (2021) demonstrate, drawing on expert

C. B. Barrett et al., *Socio-Technical Innovation Bundles for Agri-Food Systems Transformation*, Sustainable Development Goals Series,
https://doi.org/10.1007/978-3-030-88802-2_8

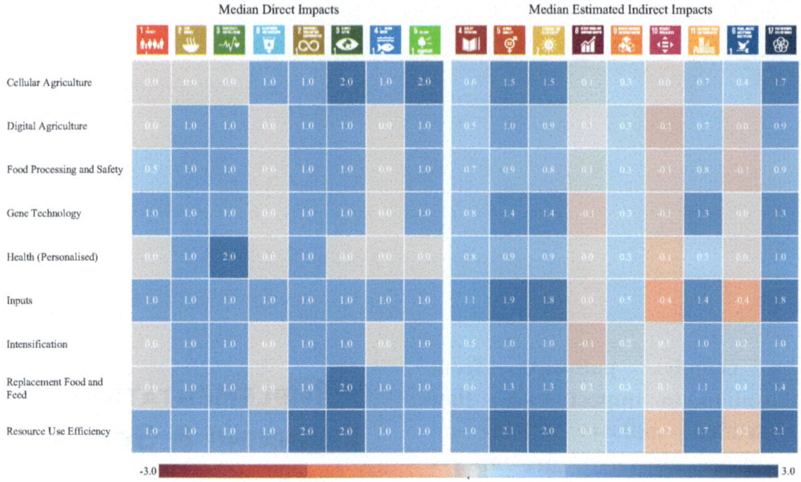

Fig. 1 Net impacts of different technology domains on food systems–related SDGs and their indirect effects on other SDGs. Indirect effects are mediated via the interactions between SDGs as quantified by Pradhan et al. (2017). Dark, mid, or light blue squares represent strong, moderate, or weak positive impacts/interactions, while grey or red squares represent neutral or negative interactions and/or impacts, respectively (Reproduced from Herrero et al. 2021)

Delphi assessments, the anticipated direct impacts of individual AFS technologies exhibit considerable heterogeneity of potential direct impacts, quite apart from the indirect effects arising from spillovers (Fig. 2). While some innovations could have very significant positive impacts, others could have neutral or negative effects on some SDGs. Additionally, while some could have highly positive impacts on one particular SDG, that same technology could exhibit a dramatically different direct impact on another SDG.

This is precisely what happened in prior eras of AFS transformation. For example, the Green Revolution introduction of improved cereals varieties, accompanied by increased irrigation and inorganic fertilizer application, elicited its desired and anticipated positive effects on staple crop productivity. But it also had less favorable direct effects on other key outcomes (e.g., water pollution) and very mixed indirect effects on still others (e.g., obesity and micronutrient deficiencies, and deforestation). The complex pathways from innovation to impact compel both broadly participatory engagement and bundling of distinct innovations, as well as careful attention paid to cultural, institutional, and policy environments that condition net impacts.

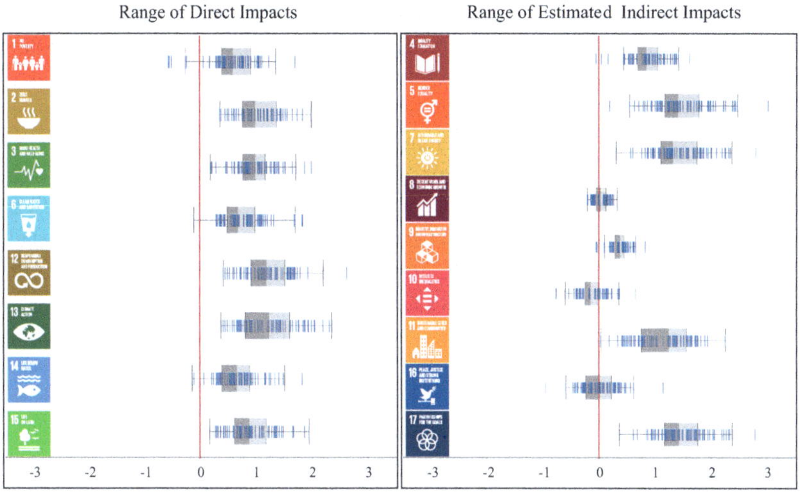

Fig. 2 Range of potential direct impacts of anticipated technologies across SDGs. Direct impacts are those that occur on the SDGs that directly relate to food systems. Indirect impacts, by contrast, are those mediated through the impacts of food systems technologies on non-food system–related SDGs. The small blue bars represent an average score of all respondents for an individual technology (Reproduced from Herrero et al. 2021)

It is therefore essential to construct mental impact pathways as part of designing transition pathways for any technological intervention in AVCs (Herrero et al. 2020, 2021). The complexity of that exercise almost inevitably requires broad stakeholder engagement and can sometimes be usefully supported by sophisticated modeling. These impact pathway mapping exercises often reveal key trade-offs and synergies arising from the multi-sectoral nature of the impacts associated with socio-technological bundles. They help identify key objectives and the indicators necessary to monitor progress towards (or away from!) those objectives. And they permit contingency planning for actions necessary to prevent or remedy undesired consequences.

Even with the best intentions, if AFS innovators focus only on direct AFS impacts, they risk adverse impacts on other, distal objectives. Because many such impacts are predictable, even if unintended, they can excite opposition to, and obstruction of, emergent technologies if a conscious effort is not made through complementary actions—a socio-technical innovation bundle—to safeguard other critical elements of human well-being. **In a pluralistic society, one must build**

coalitions of support by bundling complementary efforts that enable gains in one or more dimensions while protecting people against losses in some other dimension. The bundling strategy enables the minimization of unproductive, zero-sum contests.

As two examples, Fig. 3 shows a range of pathways to impact for two emerging, but very different, technologies: circular feeds and nitrogen fixation in cereals. These are discussed in detail above and in Herrero et al. (2021). The purpose here is not to explain each of the illustrated links—which Herrero et al. (2021) does—merely to highlight the deep interconnections that link various outcomes' responses to the introduction of even a single new technology. An innovation intended to advance progress towards one SDG inevitably generates direct and indirect effects on other SDGs as well. These spillovers are intrinsic to AFS and are one reason why bundling of social and technological innovations is so crucial to harnessing the potential of science to transform AFS.

Clearly these two technologies are being researched and sold to research agencies, venture capitalists, and the public with their intended positive impacts in mind. A more detailed examination, however, reveals positive impacts in some SDG domains—perhaps even in unintended domains—but not in others. This highlights the pervasiveness of spillover effects and the enormous difficulty of simultaneously achieving uniformly positive outcomes across SDGs. Hence the need for bundling, for broadly participatory innovation processes, and for close monitoring of a dashboard of KPMs of AFSs and beyond. We need to think not only of key accelerators and processes to ensure the adaptation and upscaling of impactful new technologies, but equally to think and plan early in the adoption and diffusion process for corrective or compensatory measures to address negative, unintended spillover effects as, and where, they emerge.

The complex pathways to impact illustrate the necessity of thinking in terms of socio-technical innovation bundles. These are necessary in the narrowest sense, to ensure the presence of socio-cultural accelerators to promote adaptation and diffusion of beneficial individual innovations, and to clear socio-cultural obstacles to upscaling. But they are, perhaps, even more important in the broader sense of addressing the inevitable spillovers and tradeoffs across diverse, desired outcomes. While innovation can often create the potential for Pareto improvements and thereby obviate the need for the typically painful zero-sum politics of food-price dilemmas—where any price change generates both winners and losers, and thus stark opposition—it still requires negotiation and compromise. Apolitical technological change does not exist. There are always winners. If navigated thoughtfully and inclusively, however, there need not be losers, just compromise to share the gains enabled by scientific and institutional advances.

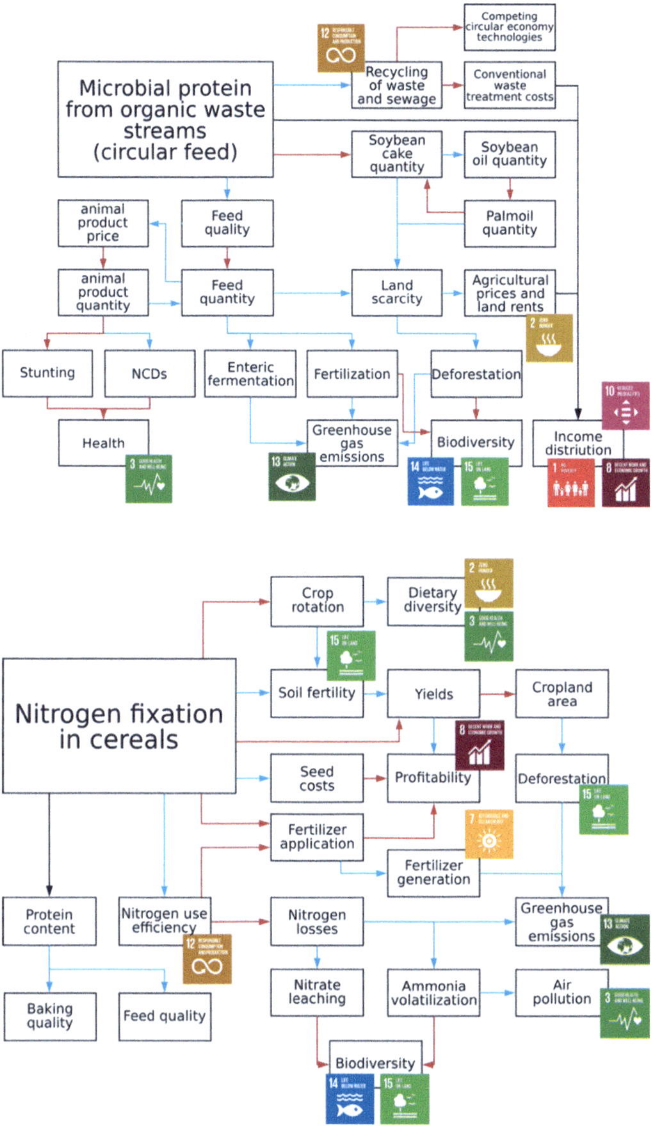

Fig. 3 Potential impact pathways of two case-study technological innovations towards the food-related SDGs: (above) personalized nutrition and (below) nitrogen-fixing cereals. Blue (red) arrows depict positive (negative) expected net impacts (Reproduced from Herrero et al. 2021)

The right socio-technical bundle, therefore, depends fundamentally on the system and on the actors involved. We cannot be overly prescriptive, as **no one-size-fits-all solutions exist**. The point, rather, is that **one must look for contextually appropriate bundles and not naively assume that an emblematic technology will automatically diffuse, much less generate favorable, intended impacts without adverse, unintended impacts**. For example, biofortification of staple crops with vitamins or minerals widely deficient in the diets of poor consumers is an extremely promising tool for advancing equitable livelihoods and healthy diets objectives in some rural and traditional systems. The crop genetic advances in biofortified germplasm often must be combined with improved agronomic practices to ensure healthy soils that deliver the needed vitamins/minerals to the re-engineered plants, with supporting seed replication and distribution systems, and with agricultural extension and nutritional education programming to inform farmers, shoppers, and caregivers of the benefits and recommended use patterns of the new variety. The precise components of biofortification-based socio-technical bundles necessarily vary, however, across crops, minerals, and AFS types. Moreover, biofortification is less useful in industrial and consolidated systems where micronutrient deficiencies have rather different etiology. The right bundle must be decided by gathering the suite of stakeholders engaged in the AFS.

References

Herrero, Mario, Philip K. Thornton, Daniel Mason-D'Croz, Jeda Palmer, Tim G. Benton, Benjamin L. Bodirsky, Jessica R. Bogard, et al. 2020. Innovation can accelerate the transition towards a sustainable food system. *Nature Food* 1 (5): 266–272. https://doi.org/10.1038/s43016-020-0074-1.

Herrero, Mario, Philip K. Thornton, Daniel Mason-D'Croz, J. Palmer, B. L. Bodirsky, P. Pradhan, C. B. Barrett, T. G. Benton, et al. 2021. Articulating the impact of food systems innovation on the Sustainable Development Goals. *The Lancet Planetary Health* 5 (1): e50–e62. https://doi.org/10.1016/S2542-5196(20)30277-1.

Pradhan, Prajal, Luís Costa, Diego Rybski, Wolfgang Lucht, and Jürgen P. Kropp. 2017. A systematic study of Sustainable Development Goal (SDG) interactions. *Earth's Future* 5 (11): 1169–1179. https://doi.org/10.1002/2017EF000632.

Towards Co-Creation of AFS Innovations by AVC Actors

A key implication of the abundance of promising technologies in various stages of development is that **AFS transformation is less likely to be limited by science-based discovery than by human agency**. What key players do in response to the wealth of options they face will ultimately determine the path(s) we follow. The key to generating, adapting, and scaling fit-for-purpose AVC innovations to advance HERS goals, therefore, turns on the coordinated exercise of human agency. The many diverse actors within value chains must both empower each other and hold all parties accountable. **Innovation co-creation is a strategic game in which each party's actions respond to others' behaviors**. Given human agency, the challenge is to structure incentives and constraints—and to identify and use KPMs necessary to monitor progress and adjust course—so as to steer actors towards mutually beneficial actions that generate

WE CANNOT BE OVERLY PRESCRIPTIVE, AS NO ONE-SIZE-FITS-ALL SOLUTIONS EXIST.

C. B. Barrett et al., *Socio-Technical Innovation Bundles for Agri-Food Systems Transformation*, Sustainable Development Goals Series,
https://doi.org/10.1007/978-3-030-88802-2_9

cooperative outcomes superior to the typically inferior outcomes that emerge from self-interested, non-cooperative behaviors.[1] **The social process of negotiating and co-creating innovation is, therefore, as important as the science of the underlying advances.**

The social process of co-creating beneficial innovation has two parts. The first is participatory dialogue and coordinated action. As we emphasized earlier, AFSs are highly decentralized, populated by myriad private actors: farmers, firm managers, chefs, food consumers, non-profit agencies, and public stewards associated with governments and civil society organizations. Each actor acts at least partly out of self-interest, guided by market signals and non-market norms and constrained by regulatory, legal, and resource limits on their actions. As we seek to accelerate the development and diffusion of beneficial innovations, it is imperative to recognize and engage with the diverse motivations people exhibit and the varied constraints they face. The task is to reconcile objectives and constraints to improve coordination and mutually reinforcing productive behaviors.

That requires dialogue. No one actor can reasonably expect (or be expected) to anticipate all the spillover effects or contingencies of a technological or institutional change in complex and heterogeneous systems. The wisdom of crowds can be tapped to improve foresight and tradeoffs analysis and to induce collective buy-in to enable progress (Surowiecki 2005). Top-down prescriptions rarely work, not even so-called expert guidance, especially in systems that demand deep contextualization and tailored bundling, much less cooperation and broad buy-in.

But the wisdom of crowds only works when no one actor has excessive, durable power, else the distorted views of the powerful persist and lead to sub-optimal outcomes (Golub and Jackson 2010). Processes that are only superficially participatory thus too often postpone—even compound—the problems that require negotiated resolution. Authentic, broad-based participation is a key reason for the impressive successes of the Science and Technology Backyards program in China, which naturally fosters innovation co-creation by researchers, farmers, input supply companies, landscape managers, etc. (see Box B).

[1] A sizable social science (especially economics) literature on "mechanism design" explores the design of optimal policy to resolve a complex collective-decision problem while accounting for both the incentives of self-interested agents and the informational and resource constraints facing each actor. The 2007 Nobel Prize in Economics was awarded to Leonid Hurwicz, Eric Maskin, and Roger Myerson largely for their contributions to mechanism design theory. Maskin's Nobel lecture offers an accessible introduction to this often-quite-technical topic (Maskin 2008).

Engaging and empowering a broad range of stakeholders is increasingly feasible with digital advances. Building public support on the demand side for beneficial innovations—among workers whose safety is imperiled, consumers whose health is compromised, farmers whose lands or livelihoods are at risk, etc.—is as important as engaging the supply side represented by researchers, policymakers, and investors. We are optimistic that as data become more plentiful and readily accessible, as the transparency of AVCs advances, and as awareness broadens and deepens of the adverse spillovers associated with current AFSs, often-marginalized elements of civil society will increasingly assert their demands. More reliable KPMs will also make it increasingly possible to identify tradeoffs and to negotiate socio-technical innovation bundles that can address multiple interests' legitimate, but disparate, concerns. Few individual innovations offer Pareto improvements; but by bundling innovations, Pareto improvements become feasible. The question is whether we have the will, and institutions, to guide us towards the imposition of strategies that reward some at the expense of others—as too often happens now—or instead towards approaches that ensure no groups get left behind. That is a sociopolitical choice.

The risk is that continued concentrated power poses grave threats to beneficial innovation. We may luck into benevolent exercise of power, whether by governments, large companies, or civil society organizations. But benevolence and an appetite for power are at best imperfectly correlated. So rules to prevent excessive concentration of power matter for innovation.

This concern about concentrated power and stakeholder access links agrifood innovation strategy to seemingly disconnected topics like the financing of political campaigns, or conflict of interest regulations to restrict civil servants' non-blinded financial interests in and hiring by private companies. The political economy of, for example, agricultural policies in the US and Europe have too often favored powerful, vested interests able to use a variety of ethically suspect maneuvers to postpone overdue reforms and discourage innovations that might threaten their short-term interests. When the powerful are the ones most likely to lose out from innovation, their natural response is to obstruct. Checks on concentrated power are essential to maintain a vibrant, innovative economy.

Hence the importance surrounding the formulation (and unending reformulation) of rules, or more broadly institutions, which Nobel Laureate Douglass North (1991) famously defined as "the humanly devised constraints that structure political, economic and social interaction." Some of the most important institutional innovations are those that rectify imbalances of power so as to amplify the voices of marginalized subpopulations and hold the relatively powerful accountable for their actions and consequences.

Concentrated power is a risk in both public and private spheres. Governments are obviously key actors in shaping AVC innovations. But states have often been slow to act and are often too easily captured by special interests. Businesses therefore play a vital role, as they may be more nimble than governments, although private firms can equally be the very special interests that capture public policy. This is perhaps especially true in AFSs, within which private businesses increasingly set food standards—related to equity, nutrition, safety, and sustainability—that are stricter than those mandated by governments, which increasingly set public standards that conform to industry-led ones (Reardon et al. 2009; Swinnen 2016; Barrett et al. in press). Firms also appear to reap most of the economic rewards from such standards (Meemken et al. 2021). This reflects growing company awareness not only of social responsibility, but equally of financial self-interest in advancing AFS innovations that enhance resilience and sustainability while promoting equity, inclusion, and healthy diets. Consumers, employees, and investors all increasingly expect firms to do more than merely maximize financial profits, and they reward them for doing so.

Publicly-funded research for development actors also has a critical role to play, and innovations are needed in the way in which research for development itself is prioritized, formulated, and implemented. Too often there are lock-ins that give primacy to priorities and practices for incremental change, and for component rather than system innovation (ISPC 2018). Much of the groundswell on the necessity of AFS transformation arises, however, from the clear need for systemic action at scale, called by different names in different contexts: "end-to-end approaches" in the development donor community, or "climate-resilient development pathways" in the climate change community, for example. We still lack comprehensive analytical frameworks associated with such approaches that can guide the co-creation of multi-dimensional, bundled actions in practice. And surely no rote, cookbook approach will ever prove feasible. Nevertheless, as we have outlined here, we know many of the elements needed. Per ISPC (2018) these include:

- New partnerships among all actors committed to moving the AFS transformation agenda forward
- Theories of change that reflect the transformational agenda of the broader design objectives—SDGs to 2030, HERS at our longer-term horizons—and that acknowledge and respect actor-specific objectives as well
- Clearer understanding of the economic, socio-cultural, and environmental dimensions of new, possibly highly disruptive technologies
- Reframing the narrative concerning how we can collectively exploit scientific advances and what new capabilities will need to be built.

The second part of the social process, following from dialogue, is coordinated actions for shared management of AFSs. **Coordination must typically be loose, in the sense of operating not through centralized decision-making but rather via structured, self-interested behavior within guardrails enshrined in institutions built through participatory dialogue, such that agents' individual, uncoordinated actions occur as if they were carefully choreographed.**[2] This is no easy task. But the to-each-his-own approach prevailing in most places in recent decades has left us on a course to climate, environmental, health, and social ruin.

The objective of the social process is to induce decentralized behaviors throughout AVCs that together drive beneficial innovations that steadily transition systems towards the HERS objectives. This requires the dialogue that is the first part. But ultimately, the key is this second part: agreed actions that together comprise a set of shared responsibilities and the KPMs and enforcement mechanisms necessary to adjust and enforce those agreements.

What are key elements of dialogues that generate the set of agreed shared responsibilities necessary to co-create beneficial socio-technical bundles? We articulate them here and then illustrate them with an example.

- **Identify system objectives as well as each actor's key incentives**. While the HERS design objectives apply in all contexts, these commonly need supplementation for a specific context. Moreover, one must not naively ignore the reality that different actors pursue varied goals: profit, political power, social or environmental outcomes, etc. Co-management works best when all parties acknowledge the diversity of stakeholder objectives. Forthrightness helps to build mechanisms that can accommodate all parties' interests. Failure to do so often leads to defection from agreements and non-cooperation, with adverse results for all parties.

---

[2] Under special (and typically unrealistic) conditions, this is the "invisible hand" of market incentives that Adam Smith (1776) so famously celebrated in his *The Wealth of Nations*, but predicated on a strong foundation of community ethical norms that Smith advanced in his earlier, foundational work, *The Theory of Moral Sentiments* (1759). Unfortunately, markets are inevitably incomplete and imperfect, so we equally need informal norms and formal laws and regulations that together induce cooperative behaviors, including compliance with formal strictures (Coase 1960; Platteau 2000; Ostrom 2010). Markets are necessary but insufficient institutions.

- **Articulate shared responsibilities**. What actions can/must different actors/organizations take in order to elicit and scale beneficial innovation in a given AFS? And what actions does each actor need others to take in order to induce these necessary actions? Identifying these shared responsibilities for mutual action is an essential step of the process.
- **Agree on key performance measures**. Organizations inevitably manage to what they measure. A manageable dashboard[3] of reliable, low-cost, transparent measures of KPMs that track progress towards goals (step 1) and success in fulfilling actors' responsibilities (step 2) is therefore essential.[4]
- **Develop open monitoring and enforcement mechanisms**. Trust is among the scarcest and most valuable renewable resources in any society (Barrett 1997; Ostrom 2010). Checks and balances are needed to prevent any actor from shirking its agreed responsibilities. But for enforcement to be credible, the consequences of failure to perform must be transparent and agreed upon.

The outcomes of these four key steps necessarily vary by food system, by technology domain, and over time. The incentives, shared responsibilities, indicators, and monitoring and enforcement mechanisms necessary for building a transformative digital stack in emerging and diversifying systems in South and Southeast Asia will surely differ considerably from those needed to guide de-agrarianization of protein production for consumers in the industrialized and consolidated systems of much of North America. Those will be different still from those needed to adapt and diffuse biofortified, nitrogen-fixing, or stress-tolerant staple-crop varieties in rural and traditional systems in sub-Saharan Africa. We cannot be prescriptive about specific actions, only about the necessity of dialogue and of agreement on shared responsibilities, KPMs, and monitoring and enforcement mechanisms.

Our panel developed one admittedly general example, abstracted from any specific system, simply to illustrate the idea. In Fig. 1, we sketch out an

[3] Given multiple objectives and responsibilities, we favor a dashboard rather than reducing multiple important objectives to a single, scalar measure through some black box index method. This is necessary so as to ensure that the introduction and scaling of innovative socio-technical bundles generate advances in at least one indicator without deterioration in any other indicator, what economists refer to as "Pareto improvement."

[4] Reliable, low-cost indicators are especially important for monetizing consumers' valuation of credence attributes (e.g., social or environmental benefits that are not directly verifiable by the consumer) and for mobilizing private capital (e.g., through impact investing) (Deutz et al. 2020; GIIN 2020).

What actions can they take?	What actions needed from others?
Set transparent standards, methods, disclosure req'ts (e.g., ISO standards), as well as agreed third-party certification processes for deviations from standards (**KPMs**: percent products covered by agreed standards)	*From researchers/industry:* capacity to automate downscaled ISO standards to product level to make affordable for SMEs (software/databases) *From governments:* regular ISO/CODEX LCA-based updates with third-party verification/cert w/LCA for deviations *From investors*: the development of impact investment instruments tied to HERS KPMs to broaden reporting buy-in
Set/enforce clear expectations on suppliers re: env't (deforestation, soils, GHGe), labor/equity (force labor, living incomes) through contract terms (**KPMs**: land in ag, median farmer/worker income, soil carbon)	*From researchers:* low cost, reliable monitoring methods *From governments:* supporting laws/enforcement, key public goods (e.g., roads/electricity/internet), sensible tax/subsidy policies, safety nets *From industry peers:* agreed standards, logos, certification/transparency methods
Reformulate/fortify/functionalize to improve product healthfulness (**KPMs**: RDA/Kcal for key micronutrients, bioactives)	*From researchers:* ID improved food science methods *From industry and governments:* agreed standards, logos, certification/transparency methods, appropriate taxes/subsidies
Build consumer awareness/valuation of HERS outcomes to generate premium value for more desirable product attributes (**KPMs**: market share of pro-HERS foods)	*From governments/media/consumer groups:* extension and popular media education on product attributes *From consumers*: learn from nutrition fact labels and FOP claims

Fig. 1 An illustrative, coordinated-action agenda regarding food product standards from the standpoint of food processors, manufacturers, and retailers. Identify distinct actors, the complementary actions each needs to take in order to generate mutually reinforcing beneficial responses, and sample key performance measures (KPMs) to track

actor-specific action agenda that food manufacturers and retailers could pursue to help drive beneficial innovations aligned with HERS objectives. This simple framework results in an enumeration of actions individual firms—and industry groups—can take that can help drive food systems in needed directions. It likewise identifies actions they need counterparties (e.g., consumer groups, governments) to take so that those firm actions are both feasible and compatible with the firm's own financial incentives. A key part of the point of the dialogues is to acknowledge each parties' legitimate, if idiosyncratic objectives. Even those private companies that authentically commit to shared societal goals, such as the SDGs or longer-run HERS objectives, must provide shareholders with satisfactory returns on equity, taking into consideration not just near-term profits, but also goodwill, market share, risk management, and stakeholder (e.g., employee, community, regulator) satisfaction goals, as well. So all parties ultimately need to identify strategies that can make good behavior sufficiently profitable to persist in the face of market pressures.

For each action, there should be an explicit KPM. The Food Systems Dashboard curates a rich list of KPMs at food system type or country level. A key to good KPMs is that they follow a standard, accepted best-measurement practice, are available reasonably universally and promptly, and are inexpensive to gather and distribute. The dashboard of KPMs appropriate to any dialogue of AFS partners will necessarily vary by system and objectives.

There exist several implications of the need for multiple parties in these social dialogues, so as to be able to agree to reciprocal action obligations. One is the necessity of building and maintaining trust and transparency. This can be difficult. It requires courageous leaders. Another is the need for adequate, locally knowledgeable scientific capacity to credibly engage on the research needs. In some places, such as much of sub-Saharan Africa, this requires collective investment in building the required, local scientific capacity where little currently exists. Third, this almost always requires cross-sector dialogue and coordination, as few of the actions and reciprocal actions needed to enable progress get confined to the boundaries of organizational charts.

Progress is feasible but fundamentally depends on AFS innovations to obviate pressing natural resource constraints and address looming food demand growth. As the Matt Ridley quote that opened the report emphasizes, "[I]nnovation is the most important fact about the modern world, but one of the least well understood" (Ridley 2020). **People too often assume that science alone can, and will, rescue us**. But translating first-rate scientific research into human progress relies inexorably on human goodwill, cooperation, and ingenuity; on incentives crafted so as to induce mutually reinforcing individual and collective behaviors; and on shared commitment both to common goals and to coordinated-action agendas that diverse stakeholders can agree on through mutually respectful dialogue. In short, human agency ultimately drives the innovation process. The most expeditious and likely path to co-creation of the socio-technical innovation bundles needed to navigate humanity away from clear and present climate, environmental, health, and socioeconomic dangers and to HERS AFSs ultimately rely on human dialogue and action. **We have met past challenges and prevailed. We can do so again. But we cannot afford to delay**.

References

Barrett, Christopher. 1997. Idea gaps, object gaps, and trust gaps in economic development. *Journal of Developing Areas* 31 (4): 553.

Barrett, Christopher B., Thomas Reardon, Johan Swinnen, and David Zilberman. In press. Agri-food value chain revolutions in low- and middle-income countries. *Journal of Economic Literature*.

Coase, Ronald H. 1960. The problem of social cost. In *Classic papers in natural resource economics*, 87–137. London: Palgrave Macmillan.

Deutz, Andrew, Geoffrey M. Heal, Rose Niu, Eric Swanson, Terry Townshend, Li Zhu, Alejandro Delmar, Alqayam Meghji, Suresh A. Sethi, and John Tobin-de la Puente. 2020. *Financing nature: Closing the global biodiversity financing gap*. The Paulson Institute, The Nature Conservancy, and the Cornell Atkinson Center for Sustainability.

Global Impact Investing Network (GIIN). 2020. *Annual impact investor survey 2020*. https://thegiin.org/assets/GIINpercent20Annualpercent20Impactpercent20Investorpercent20Surveypercent202020.pdf.

Golub, Benjamin, and Matthew O. Jackson. 2010. Naive learning in social networks and the wisdom of crowds. *American Economic Journal: Microeconomics* 2 (1): 112–149. https://doi.org/10.1257/mic.2.1.112.

ISPC. 2018. Public agricultural research and development in an era of transformation. Volume 1—Analysis and reflections. Rome: CGIAR Independent Science and Partnership Council (ISPC).

Maskin, Eric S. 2008. Mechanism design: How to implement social goals. *American Economic Review* 98 (3): 567–576. https://doi.org/10.1257/aer.98.1.567.

Meemken, Eva-Marie, Christopher B. Barrett, Hope C. Michelson, Matin Qaim, Thomas Reardon, and Jorge Sellare. 2021. Sustainability standards in global agrifood supply chains. *Nature Food* (2): 758–765. https://doi.org/10.1038/s43016-021-00360-3.

North, Douglass C. 1991. Institutions. *Journal of Economic Perspectives* 5 (1): 97–112.

Ostrom, Elinor. 2010. Beyond markets and states: Polycentric governance of complex economic systems. *The American Economic Review* 100 (3): 641–672.

Platteau, Jean-Philippe. 2000. *Institutions, social norms, and economic development*. Amsterdam, The Netherlands: Harwood Academic Publishers.

Reardon, Thomas, Christopher B. Barrett, Julio A. Berdegué, and Johan F.M. Swinnen. 2009. Agrifood industry transformation and small farmers in developing countries. *World Development* 37 (11): 1717–1727. https://doi.org/10.1016/j.worlddev.2008.08.023.

Ridley, Matt. 2020. *How innovation works*. New York: Harper.

Smith, Adam. 1759. *The theory of moral sentiments*. HG Bohn.

Smith, Adam. 1776. *The wealth of nations*. London: W. Strahan and T. Cadell.

Surowiecki, James. 2005. *The wisdom of crowds*. New York: Anchor.

Swinnen, Johan. 2016. Economics and politics of food standards, trade, and development. *Agricultural Economics* 47 (November): 7–19.

Technical Appendix

Food Systems Typologies[1]

As detailed in Marshall et al. (2021), the food systems typologies we and the Food Systems Dashboard use were developed using a method that began with a structured scoping review of the existing food systems literature. Based on the variables found in that review, a parsimonious set of four variables was identified from which to build the typology using simple quantitative methods. The variables were selected according to the following criteria: (1) the group of indicators chosen should reflect different components of the food system; (2) the literature should support the indicators' association with food system patterns and transitions; and (3) indicators should have high global coverage, including across different regions and income group classifications.

The four component indicators are: agricultural value-added per worker; the percent of dietary energy from cereals, roots, and tubers (i.e., staple foods); the number of supermarkets per 100,000 inhabitants; and urban population as a percent of total population. The underlying data were sourced from the World Bank, FAO, Food Balance Sheets, and Euromonitor. We include 155 countries for which data was available for all four indicators.

For each indicator, countries were ranked from highest to lowest, under the hypothesis that higher values were associated with more "modern" food systems, and lower values with more "traditional" food systems. The ranking was inverted

[1] These details are adapted from Marshall et al. (2021), which provides considerably more detail on the method, its validation, and application.

C. B. Barrett et al., *Socio-Technical Innovation Bundles for Agri-Food Systems Transformation*, Sustainable Development Goals Series,
https://doi.org/10.1007/978-3-030-88802-2_10

in the case of the share of dietary energy from cereals, roots, and tubers, which is theorized to decrease as food systems grow more advanced.

Each country was assigned a score equal to the sum of its ranks on each of the four indicators. For example, if a country ranked tenth on agricultural value-added; fifteenth on share of dietary energy from cereals, roots, and tubers; seventeenth on number of supermarkets per 100,000 population; and eighth on urbanization, its score was 50. After calculating scores for each country, we sorted countries from lowest to highest score and divided them into quintiles. The lowest quintile represents the most modern AFS type and the highest the most traditional system type. Cross-system patterns in the four underlying variables used to create the typology align with narratives provided by the food systems, nutrition transition, and structural transformation literatures, exhibiting statistically significant variation among groups even though substantial heterogeneity still exists within individual food system types.

Country classifications based on this method were then validated against a host of other food systems–related indicators that were not used to construct the rankings and the typologies. We found that the typology, indeed, generates sensible patterns in a broad suite of diet, nutrition, health, socioeconomic, and environmental outcomes across the resulting food system types.

Innovations' Impacts on SDGs[2]

Herrero et al. (2020, 2021) collated an inventory of future technologies that could accelerate progress towards achieving the food systems–related SDGs. To assemble the inventory of possible technologies, they carried out literature searches around the idea of AFS transformation. The literature searches were complemented by researcher expert opinions. They found approximately 80 technologies that directly addressed some dimension of the food system. These were classified into the following technology groups: cellular agriculture, digital agriculture, food processing and safety, gene technology, inputs, intensification, replacement food and feed, and others.

The key criteria for inclusion in the inventory were that (1) technologies needed to have a direct impact on the key processes associated with the food

[2] These details are adapted from Herrero et al. (2021), which provides considerably more detail on the methods used for data generation and analysis, as well as additional explanation of the these and other specific innovations analyzed.

system from production to consumption, and (2) they represented "products" of some type that were applicable to the food system. The inventory contains some groupings of very similar technologies for which it would be difficult to separate the magnitudes and types of their impacts. Many management and system-level interventions, as well as technologies from other sectors, will undoubtedly also play critical roles in improving food systems, but these are not specifically covered here (Herrero et al. 2020).

Herrero et al. (2020) classified each technology by its Technology Readiness Index and elicited expected qualitative assessments of its potential impacts and likelihood of adoption to 2030 (to align with the SDG time horizon) using an online ranking tool to evaluate and score each of the technologies with respect to three characteristics. The potential impact of the technology on the SDGs is of particular relevance to this report. The most directly AFS-relevant SDGs for which technology's potential impact was directly elicited were the following:

- SDG 1: End poverty in all its forms everywhere
- SDG 2: End hunger, achieve food security and improved nutrition, and promote sustainable agriculture
- SDG 3: Ensure healthy lives and promote well-being for all at all ages
- SDG 6: Ensure availability and sustainable management of water and sanitation for all
- SDG 12: Ensure sustainable consumption and production patterns
- SDG 13: Take urgent action to combat climate change and its impacts
- SDG 14: Conserve and sustainably use the oceans, seas, and marine resources for sustainable development
- SDG 15: Protect, restore, and promote sustainable use of terrestrial ecosystems, sustainably manage forests, combat desertification, halt and reverse land degradation, and halt biodiversity loss.

The potential impacts were scored using a 7-point score as follows:

3: A large positive impact
2: A moderate positive impact
1: A small positive impact
0: A neutral (neither positive nor negative) impact
−1: A small negative impact.
−2: A moderate negative impact
−3: A large negative impact

The survey was completed by 32 experienced respondents, all authors of Herrero et al. (2020). Respondents were asked to score only those technologies for which they felt comfortable providing an opinion. Descriptive statistics were then calculated for each technology and the results analyzed statistically. For more details, see Herrero et al. (2020, 2021).

References

Herrero, Mario, Philip K. Thornton, Daniel Mason-D'Croz, Jeda Palmer, Tim G. Benton, Benjamin L. Bodirsky, Jessica R. Bogard, et al. 2020. Innovation can accelerate the transition towards a sustainable food system. *Nature Food* 1 (5): 266–272. https://doi.org/10.1038/s43016-020-0074-1.

Herrero, Mario, Philip K. Thornton, Daniel Mason-D'Croz, J. Palmer, B. L. Bodirsky, P. Pradhan, C. B. Barrett, T. G. Benton, et al. 2021. Articulating the impact of food systems innovation on the Sustainable Development Goals. *The Lancet Planetary Health* 5 (1): e50–e62. https://doi.org/10.1016/S2542-5196(20)30277-1.

Marshall, Quinn, Jessica Fanzo, Christopher B. Barrett, Andrew D. Jones, Anna Herforth, and Rebecca McLaren. 2021. Building a global food systems typology: A new tool for reducing complexity in food systems analysis. *Frontiers in Sustainable Food Systems* 5: 432. https://doi.org/10.3389/fsufs.2021.746512.

Index

A
Africa, 25, 49–51, 60, 65, 79, 80, 86, 89, 90, 107, 108, 113, 114, 122, 159
Agri-food systems (AFS), 1–12, 14–17, 21, 22, 25, 30, 32, 34–37, 39, 47, 49–51, 53, 55, 56, 58–61, 65–67, 69–71, 73–77, 79, 80, 82–87, 93, 97, 99, 100, 108, 109, 114, 126, 129, 133, 137, 160, 161, 165–167, 170–172, 174, 178–182, 184, 188
Agrochemicals, 15, 22, 28, 29, 31, 87, 101, 102, 110, 112–114, 117, 119, 134, 170
Agroecology/ical, 12, 58, 61, 106, 109–113, 117
Alternative proteins
  cellular protein, 129, 134
  plant-based protein(s), 129
  tissue engineering, 52, 120
Aquaculture, 9, 52, 77, 103, 108, 109, 117–120, 189
Asia, 25, 48, 49, 57, 86, 122, 160
Australia, 136

B
Biodiversity, 14, 29, 31, 35, 37, 39, 52, 61, 70, 71, 87, 88, 92, 112, 113, 129, 166, 189
Bundle/bundling, 4–10, 16, 49, 55, 65, 79, 85, 98, 160, 162, 167, 169–172, 174, 178–182, 185

C
Calories, 26, 28, 71
Cereals, 13, 21, 22, 55–58, 66, 69, 83, 84, 100–105, 110, 111, 117, 118, 129, 134, 135, 159, 161–163, 170, 172, 173, 187, 188
Certification, 52, 82, 90, 93, 124–126
Circular economy, 74, 78, 109
  circular feeds, 52, 133, 165, 172
Climate
  carbon, 14, 28, 29, 52, 65, 83, 104, 114, 115, 119, 122, 124, 129
  change, 4, 5, 24, 29, 30, 35–38, 51, 61, 65, 71, 103, 108, 115, 133, 189
  fossil fuel(s), 15, 28, 29, 52, 109
  greenhouse gas(es), 14, 15, 28, 29, 38, 48, 51, 52, 61, 62, 70, 90, 92, 103, 107, 108, 114, 115, 122–124, 128
  variability, 138
  weather, 2, 79, 81, 82, 85
Co-design, 7, 9
Conflict, 2, 11, 25, 30, 38, 56, 61, 93, 97, 179
Controlled environment agriculture (CEA), 52, 74, 78, 121, 122

C. B. Barrett et al., *Socio-Technical Innovation Bundles for Agri-Food Systems Transformation,* Sustainable Development Goals Series,
https://doi.org/10.1007/978-3-030-88802-2

COVID-19, 3, 9, 25, 30, 32–39, 52, 55, 60, 67, 70, 79, 93, 133
Crops, 22, 24, 30, 32, 34, 56, 58, 60, 70, 81, 83, 91, 101, 102, 104, 107, 109–115, 117, 134, 159, 160, 170, 174, 182
Cuisine, 7

D
De-agrarianize/ation, 9, 121–123, 133
Deconcentrate, 9, 70, 71, 164
Deforestation, 15, 28, 111, 170
Diet(s)
  dietary preferences, 59
  healthy, 2, 24, 25, 33, 37, 38, 58–60, 69, 71, 75, 89, 93, 96, 97, 117, 122, 133, 136, 137, 174, 180
  unhealthy, 27, 59, 87, 92, 135, 137
Digital technologies
  computers/ing, 83, 133
  connectivity, 16, 78–80, 84, 98
  database(s), 84, 85
  e-commerce, 34
  internet, 80
  machine learning, 74, 80, 81, 83, 93, 99, 116, 137
  mobile phone(s), 79
  precision agriculture, 81, 83
  robot(s, ic, ics), 34, 74, 83, 85
  sensors, 79, 81, 83, 85, 99, 137
  stack, 80, 182
  transfers, 9, 35, 87, 94–96, 101, 103, 104, 123, 163, 166
Disasters, 2–4, 8, 9, 12, 26, 29–39, 55, 58, 60, 61, 65, 70, 71, 75, 80, 85, 88–90, 93, 95–97, 105, 111, 117, 122, 131, 133, 171, 179, 180, 184
Disease
  animal, 7, 13–15, 28, 30, 32, 33, 39, 51, 52, 56, 61, 66, 78, 79, 81, 100, 101, 104, 105, 108, 109, 111, 115, 118–121, 128, 131, 134, 161, 163, 165
  human, 29, 39, 79, 120, 163
  infectious, 34, 35, 60, 61
  non-communicable, 35, 67, 96, 107, 135
  plant, 7, 15, 28, 56, 61, 79, 81, 85, 100, 101, 104, 109, 111, 128

E
Early warning systems, 30
Engagement, 7, 77, 80, 97–99, 106, 112, 170, 171
Equity, 4, 9, 16, 24, 25, 35–37, 56, 59, 60, 89, 93, 97, 98, 109, 112, 119, 122, 128, 133, 169, 180, 184, 188
Europe, 32, 47, 48, 113, 179
Extensification, 56, 61
Extension, 10, 57, 69, 79–81, 92, 106, 107, 174
Extinction, 2, 9, 35, 52, 53, 87

F
Famine, 25, 30, 55, 57
Farms, 2, 12–15, 34, 56, 60, 61, 79, 80, 82, 88, 109, 111, 112, 121–123
Fermentation, 52, 109, 121, 122, 129, 131
Fertilizer, 13, 15, 28, 29, 31, 57, 81, 87, 101, 103, 106, 112, 114–117, 131, 165
  chemical, 57, 65, 71, 110, 114, 116, 170
  organic, 31, 115, 116, 131
Finance, 3, 8–10, 14, 16, 17, 30, 31, 36–39, 69, 74, 75, 77, 79, 83, 85–95, 99, 101, 107, 120–122, 159, 160, 164, 166, 167, 169, 179, 180, 184
Firms, 5, 12, 14–17, 27, 32, 34, 49, 74, 75, 80, 81, 87–89, 92, 121, 124, 178, 180, 184
Food
  assistance, 33, 35, 94–97, 118, 167, 168
  demand, 2, 9, 13, 33, 49–52, 61, 86, 89, 184
  Food Systems Dashboard, 12, 184, 187
  label, 126
  loss, 12, 13, 16, 27, 28, 34, 131, 133
  package/ing, 13, 66, 126
  price(s), 16, 17, 25, 27, 56, 87, 97, 99, 132, 172

production, 3, 14, 16, 30, 34, 52, 61, 78, 102, 105, 115, 117, 122, 128
regulation, 39, 137
safety, 10, 13, 39, 81, 99, 117, 121, 136, 137, 163
security, 2, 11, 28, 59, 86, 95, 108, 189
standards, 13, 14, 59, 126, 180
supply, 13, 30, 66, 97, 133, 137
systems, 13–15, 21, 23, 30, 52, 79, 81, 84, 105, 107, 110, 112, 117, 125, 135, 162, 169–171, 182, 184, 187–189
trade, 12–14, 32, 96, 99
waste, 13, 27, 84, 99, 115, 131, 133
Fortification
biofortification, 102, 134, 167, 174
industrial, 14, 21, 52, 115, 118, 119, 135, 174
salt, 14, 15, 59, 102, 136, 168
Fruits, 13, 14, 21, 32, 33, 52, 70, 82, 84, 100, 104, 105, 121, 129, 131, 135, 137, 166, 167

G
Genetic
advances, 5, 69, 100, 107, 133, 167, 174
breeding, 69, 100–102, 104, 105, 107, 111
engineering, 69, 101, 102, 105, 112, 133, 163
genomic, 107
germplasm, 101, 105, 174
hybrid, 101
Globalization, 38, 65
Golden rice, 69, 102, 134, 161
Government(s), 3, 7, 9, 10, 13, 14, 16, 32, 33, 35, 37, 38, 57, 79, 80, 82, 86, 87, 89, 91, 97, 99, 107, 114, 122, 124, 125, 130, 133, 136, 161, 164, 178–180, 184
Green Revolution, 55–57, 86, 87, 160, 161, 170

H
Heterogeneity, 4, 12, 17, 31, 77, 89, 126, 170, 188
Human agency, 5, 7, 9, 11, 12, 162, 177, 185

I
Income, 2, 5, 12–14, 22, 24–26, 28, 29, 33, 35–37, 49–53, 58–60, 65, 67, 69, 71, 75, 77, 81, 86, 89, 91, 92, 94, 99, 100, 105, 107, 108, 118, 119, 121, 122, 131, 137, 138, 174, 179, 187
Inequality, 59, 87, 100, 115
Infrastructure, 10, 13, 51, 58, 79, 81, 87, 123, 125, 130–133, 138, 161, 164
Insects, 77, 101–103, 108–110, 112, 117, 118
Intellectual property rights, 9, 14, 24, 36, 69, 85, 91, 92, 161
Intensification, 28, 61, 106, 107, 111, 113, 117, 188
Irrigation, 22, 28, 91, 161, 170

K
Key performance metrics, 6, 7, 9, 124, 125, 128, 172, 177, 179, 181–184

L
Labor, 2, 13, 22–25, 32, 56, 57, 59, 60, 66, 81, 94, 95, 99, 166
Land, 2, 7, 9, 13–15, 22–24, 29, 31, 39, 47, 52, 56, 59–62, 65, 71, 78, 109, 115–117, 121–123, 129, 132, 133, 165, 166, 189
Latin America and Caribbean, 48, 57, 60, 86, 108
Livestock, 15, 21, 31, 33, 34, 51, 52, 56, 59, 77, 78, 80, 82, 83, 103, 107, 108, 110, 111, 117–122, 133, 159, 162, 166

M

Macronutrients
calorie(s), 26, 55
dietary energy, 25, 55, 187, 188
Fat(s), 14, 15, 26, 27, 59, 119, 135, 136
protein(s), 28, 51, 52, 78, 108, 117–121, 134
Malnutrition
hunger, 30
obesity, 26, 65–67, 135
stunting, 65, 66
wasting, 66
Micronutrients
deficiencies, 13, 28, 99, 114, 134, 135, 167, 170, 174
minerals, 25, 28, 102, 103, 108, 114, 120, 134, 135, 167, 174
vitamins, 25, 28, 102, 103, 108, 134, 135, 167, 168, 174
Mycotoxin(s), 101, 117, 119, 129

N

North America, 48, 182
Nutrition, 1, 28, 65–67, 77, 81, 85, 86, 101–104, 107, 108, 119–122, 124, 126–129, 131, 133–138, 168, 173, 174, 180, 188, 189

O

Open-source, 9, 36
Organic, 29, 110, 114, 115, 118, 119, 130, 131, 163

P

Palatability, 26
Pareto, 8, 12, 16, 31, 59, 67, 70, 87, 100–102, 104, 105, 110, 113, 125, 167, 172, 179
Pest(s), 30, 31, 52, 78, 81, 92, 101–103, 105, 109, 110, 112–114, 134
Philanthropy, 3, 9, 17, 57, 74, 86, 87, 92, 161
Political economy, 5, 179
Population growth, 14, 25, 49–51, 77
Poverty, 22, 25, 58, 59, 93, 96, 111, 189
Processing, 2, 13–15, 24, 26, 27, 30, 33, 37, 38, 51, 59, 66, 67, 69, 74, 77, 79, 92, 99, 116, 128, 129, 131, 133–136, 164, 165, 188
Productivity, 3, 4, 8, 13, 21–25, 31, 49, 53, 56–61, 69, 81, 89, 100, 107, 108, 111, 114, 117, 170

R

Reformulation, 96, 134–136, 179
Research and development, 21, 24, 36, 69, 70, 74, 86, 87, 89, 91–93, 98, 159, 162
Resilience, 2–4, 29–31, 34, 35, 37, 58, 60, 69, 71, 75, 85, 105, 111, 113, 122, 131, 180
Restaurants, 2, 7, 14, 27, 33, 34, 38, 51, 74
Roots and tubers, 21, 102, 103, 110, 112
Rural, 10, 12–15, 21, 22, 24, 28, 47, 49, 52, 59, 60, 65, 66, 79, 80, 82, 87, 94, 95, 98, 107, 111, 115, 116, 122, 123, 137, 167, 174, 182

S

Salinization, 29
Scaling, 6, 9, 77, 119, 162, 177
Science and Technology Backyard (STB), 106, 107, 112, 178
Social justice, 1, 8, 21, 86, 88
Social protection, 10, 30, 33, 37, 60, 77, 93–97
Soil
biology, 116
chemistry, 116
degradation, 29, 71
health, 15, 71, 114–117, 131, 174
nutrients, 31, 102, 110, 114
organic matter, 114
physics, 116
South America, 114

Spillovers, 1, 4, 8, 9, 11, 12, 15–17, 21, 38, 57, 60, 71, 86–88, 92, 100, 125, 128, 160, 165, 169, 170, 172, 178, 179
Sub-Saharan Africa, 22, 24, 47, 48, 86, 161, 182, 184
Subsidies, 10, 16, 33, 71, 87, 92–94, 96, 98, 100, 123, 137
Supermarkets, 2, 13, 14, 34, 35, 51, 67, 68, 82, 84, 187, 188
Sustainability, 4, 24, 28, 34, 35, 37, 38, 52, 58, 61, 69, 71, 75, 81, 85, 89, 105, 110, 117, 122–124, 128, 131, 134, 136, 180
Sustainable Development Goals (SDGs), 4, 10, 11, 49, 58, 59, 61, 75, 77, 115, 124, 125, 128, 164, 166, 169–173, 180, 184, 188, 189

T

Taxes, 92, 122, 136, 137
Technology
  change, 8, 24, 66, 77, 82, 93, 98, 101, 105, 106, 122, 133, 160, 172, 178
  innovation, 1, 4, 7, 8, 16, 28, 49, 74, 75, 77, 79, 83, 85, 91, 93, 98, 100, 105, 107, 117, 131, 137, 160, 162, 164, 167, 170–173, 178
  leapfrogging, 122
  readiness, 75–77, 120, 137, 159, 189
Traceability, 38, 82, 125, 128, 131, 137
Trust, 7, 35, 38, 69, 70, 80, 106, 125, 126, 128, 162, 163, 182, 184
Typologies, 14, 15, 65, 66, 187, 188

U

Undernourishment, 25, 26, 96
Undernutrition, 3, 65, 96
Universal Declaration of Human Rights, 12, 59
Urban, 13, 14, 22, 47, 52, 60, 65, 111, 116, 121, 122, 128, 131, 187

V

Value chains, 2–4, 7, 8, 11–14, 17, 21, 22, 24, 25, 29–34, 37–39, 47–49, 51, 52, 59, 65, 70, 74, 75, 77, 82–84, 87, 105, 108, 115, 117, 118, 121, 122, 124–126, 128–131, 133, 136, 162, 163, 171, 177, 179, 181
Vegetables, 13, 21, 28, 32, 33, 52, 82, 100, 105, 129, 131, 135, 137, 167

W

Water
  irrigation, 57
  pollution, 115, 170
  shortages, 105
Weeds (weeding), 31, 83, 101, 103, 110, 112

Y

Yields, 12, 21, 22, 24, 53, 57, 58, 60, 61, 80, 101, 102, 104, 106, 107, 110–113, 118, 159

Z

Zoonose, 39, 52, 61, 166

The manufacturer's authorised representative in the EU is Springer Nature Customer Service Centre GmbH, Europaplatz 3, 69115 Heidelberg, Germany. If you have any concerns regarding our products, please contact ProductSafety@springernature.com

Printed and bound by CPI Group (UK) Ltd, Croydon, CR0 4YY
30/09/2026
02225456-0002